Springer Aerospace Technology

The series explores the technology and the science related to the aircraft and spacecraft including concept, design, assembly, control and maintenance. The topics cover aircraft, missiles, space vehicles, aircraft engines and propulsion units. The volumes of the series present the fundamentals, the applications and the advances in all the fields related to aerospace engineering, including:

- structural analysis,
- aerodynamics,
- aeroelasticity,
- aeroacoustics,
- flight mechanics and dynamics
- orbital maneuvers,
- avionics,
- systems design,
- materials technology,
- launch technology,
- payload and satellite technology,
- space industry, medicine and biology.

The series' scope includes monographs, professional books, advanced textbooks, as well as selected contributions from specialized conferences and workshops.

The volumes of the series are single-blind peer-reviewed.

To submit a proposal or request further information, please contact: Mr. Pierpaolo Riva at pierpaolo.riva@springer.com (Europe and Americas) Mr. Mengchu Huang at mengchu.huang@springer.com (China)

The series is indexed in Scopus and Compendex

More information about this series at http://www.springer.com/series/8613

Oleg Ivanovich Zavalishin ·
Dmitry Alexandrovich Zatuchny ·
Yury Grigorievich Shatrakov

Modern Requirements for Noise Immunity Aircraft Navigation Equipment

Springer

Oleg Ivanovich Zavalishin
Moscow, Russia

Dmitry Alexandrovich Zatuchny
Moscow, Russia

Yury Grigorievich Shatrakov
Saint Petersburg, Russia

ISSN 1869-1730 ISSN 1869-1749 (electronic)
Springer Aerospace Technology
ISBN 978-981-16-0072-2 ISBN 978-981-16-0073-9 (eBook)
https://doi.org/10.1007/978-981-16-0073-9

This Springer imprint is published by the registered company Springer Nature Singapore Pte Ltd.
The registered company address is: 152 Beach Road, #21-01/04 Gateway East, Singapore 189721, Singapore

Introduction

At the end of the 20th century, the long-held dream of aeronauts to be able to navigate around the clock at any point in the world's airspace through a single "seamless" through-system came true. The global navigation satellite systems (GNSS) GLONASS and GPS have become such systems in civil aviation. These systems are standardized by the International Civil Aviation Organization (ICAO) and, despite the huge costly ground infrastructure, civil signals are available to users free of charge (the declarations of the head of state—the owner of the system).

In the materials of the 12th Air Navigation Conference of ICAO, the satellite navigation is called the "cornerstone" in the navigation development of world civil aviation.

At the same time, the early period of the implementation of GNSS technology in civil aviation showed a number of specific problems connected with the uneven distribution of radio signals in time and space and other adverse factors.

These problems become significant at the stages of the flight of aircraft requiring a limited time resource for making decisions and taking actions by the aircraft crew (AC)—instrumental precision AC approach and landing. Ignoring the errors connected with abnormal influences on GNSS signals when using satellite navigation in civil aviation, especially as a primary facility, can lead to dangerous incidents and even to air accidents.

The most unpredictable factors of decrease in the accuracy of navigation are ionospheric disturbances (magnetic storms) under the influence of space and, in particular, solar photon radiation. Up to the present moment, there are no adequate models of magnetic storms in the world and, as a result, models of delayed radio signals estimation of the SRNS in the earth's ionosphere. The existing CONUS and Plasma Bubble models are very conditional, since the sample is small and unrepresentative. The model of nuclear reactions at the sun (protuberances) does not exist either. All these phenomena cannot be taken into account by the on-board navigation user equipment (NUE), leading to the violation of the integrity of the determination of the coordinates of aircraft.

Integrity is understood as a confidence-building measure—the inverse value of the probability of an undetected failure (provided that the alarm is triggered within the time specified by the ICAO standards).

A rare (but unpredictable) integrity violation factor is the pseudo random sequence (PRS) distortion at the satellite signal generator, which cannot be detected by on-board navigation user equipment due to its relative simplicity. The detection of such effects is called SQM. In addition, local jamming conditions in the aerodrome area have a significant impact on the quality of satellite navigation. The best way to deal with undetected errors is the measurement (and not the calculation) of pseudo range errors in the area where the landing or landing approach is performed in real time and their immediate (not later than in 1 sec) transmission to the on-board navigation equipment.

The foreign and Russian studies, as well as analysis of GLONASS and GPS interface control documents, showed that the confidence-building measure in GLONASS and GPS signals does not meet ICAO's integrity requirements for satellite signal in space during flight stages connected with AC terminal procedures, landing approaches and landing. The ICAO expert groups have determined that to use the GLONASS and GPS SRNS as the primary means of navigation, it is necessary to increase its integrity through the use of functional augmentations.

Improving the noise immunity of the navigation systems of civil aviation aircraft is one of the necessary conditions for meeting the ICAO integrity requirements. It should be noted that in the development and operation of tools that increase noise immunity, in addition to the noise situation, it is necessary to take into account the type of aircraft, the specifics of the regions over which the flight course lies and the stages of flight, especially those that require a limited time resource.

This paper proposes the practical ways to improve the noise immunity of the navigation systems of civil aircraft for various cases.

Contents

Abbreviations

AA	Assembly average
AC	Aircraft
ACL	Aircraft laboratory
ACSU	Automatic control system unit
ADC/DAC	Analog-to-digital and digital-to-analog converters
AFD	Antenna-feeder device
AIP	Aeronautical information publication
ARINC	International Common Standard for Civil Aviation
ATC	Air traffic control
ATIS	Automatic terminal information service
ATM	Air traffic management
AV	Air vehicle
AVS	Automatic vision system
AWC	Adverse weather conditions
AWOG	The group of experts of the European Bureau of ICAO on all-weather flights
BR	Base receiver
CA	Civil aviation
CAT	Category
CAS CCIM	Complex automated system for collecting and communicating to aviation users in the airspace of the Russian Federation of information on the status of orbital groupings of the global navigation satellite system and means of functional augmentations
CCC	Command and control centre
CNS/ATM	Communications, navigation, surveillance/air traffic management
CONUS	Contiguous United States (ionospheric storm model)
CSRTI	Central Scientific and Research Testing Institute
CSSU	Computing and switching system unit
CTS	Complex technical systems
DB	Database
DBMS	Data base management system
DC	Differential corrections

DDRE	Differential data reception equipment
DGNSS	Differential global navigation satellite system
DL	Datalink
DM	Differential mode
DQM	Data quality monitor (satellite data estimation method)
DT	Desired track
DTS	Delay tracking schemes
EGNOS	European Geostationary Navigation Overlay Service
EP	Earth parameters
ETS	Ephemerical-time support
EVS	Enhanced vision system
FAA	Federal Aviation Administration
FAR	Federal Aviation Regulations
FAS	Final approach segment
FBX	Filmbox (file format)
FD	Flight director
FLS	Flight landing strip
FNI	Flight navigation instrument
FTE	Flight technical error
GAD	Ground accuracy designator (GBAS system accuracy class)
GAST	DGBAS Approach Service Type D (the category to landing approach and landing assurance according to the ICAO requirements)
GASTC	GBAS Approach Service Type C (the category to landing approach and landing assurance according to the ICAO requirements)
GBAS	Ground-based augmentation systems (local functional extension to ground-based GNSS)
GFTC	Ground and flight test complex
GLONASS	Global navigation satellite system of Russia
GNSS	Global navigation satellite system
GOST	State standard
GPS	Global positioning system (USA)
GSLT	Glide slope localizer transmitter
GSLTS	Glide slope localizer transmitter system
GSS	Glide slope system
HAC	Higher Attestation Commission (under the Ministry of Education and Science of the Russian Federation)
IAC	Information Analysis Centre
IC	Integrity control
ICAO	International Civil Aviation Organization, the specialized UN agency
ILS	Instrument landing system
INLS	Integrated navigation and landing system
INS	Inertia navigation system
IONO	Ionosphere

IR	Infrared
LMCS	Local monitoring and correcting station (GBAS)
LPL	Lateral protection level
LSM	Least square method
MC	Multi-constellation category
MF	Multi-frequency
MFI	Multi-function indicator
MLMCS	Mobile local monitoring and correcting station
MLS	Microwave landing system
MQM	Measurement quality monitoring
MSAS	Multi-functional satellite augmentation system (Japan)
MTBF	Mean time between failures (for one subsystem)
MTBW	Mean time between breaks in work
NAGU	Notice advisory to GLONASS
NANU Notice	Advisory to Navstar users (GPS status message)
NOTAM	Notice to airmen
NPA	Non-precision approach
NS	Navigation satellite
NSC	Navigation spacecraft
NSE	Navigation system error
NT	Navigation task
NUE	Navigation user equipment
OES	Optoelectronic system
OMS-L	On-board multipurpose system (landing)
OTMC	On-board trajectory management complex
PC	Phase corrections
PDE	Path definition error
PEE	Position estimation error
PP	Pseudo phase
PPC	Pseudo phase combinations
PR	Pseudo range
PRC	Pseudo range correction
PRS	Pseudo random sequence
PZ 90.02	GLONASS coordinate system
RAIM	Receiver autonomous integrity monitoring
RMD	Reception measuring device
RMS	Root mean square
RMSD	Root-mean-square deviation
RNAV	Area navigation
RNP	Required navigation performance
RTCA	Radio Technical Commission for Aeronautics
RTF	Radio-technical facilities
RTFS	Radio-technical flight support
RVR	Runway visual range
SARP's	Standards and recommended practices of ICAO

SBAS	Satellite-based augmentation system
SDCM	System for differential corrections and monitoring
SIS	Signal in space
SLS	Satellite landing system
SNR	Signal-to-noise ratio
SNS	Satellite navigation system
SP	Smoothed phases (of measurements)
SQM	Signal quality monitoring
SR	Satellite receiver
SRNS	Satellite radio navigation system
SRU	Satellite receiver unit
SS	Scientific software
STM	Signal time mark
SVS	Synthetic vision system
TI	Thermal imaging
TS	Time scale
TSE	Total system error
UE	User equipment
UV	Ultraviolet
UV-B	Ultraviolet-blind
VDB	Very high frequency data broadcast
VHF	Very high frequency
VPL	Vertical protection limit
WAAS	Wide area augmentation system (USA)
WGS	World geodetic system
WGS-84	World geodetic system, GPS coordinate system

Chapter 1
The Method of Aircraft Landing with the Use of Integrated Satellite Optical Navigation System

1.1 A Comparative Analysis of Existing Instrumental Landing Systems that Meet the Current ICAO Requirements, and the Rationale for the Expediency of the Transition to the Aircraft Landing Using Satellite Radio Navigation Systems. The Methods for Compensation of Pseudo Range Measurement Errors in Satellite Radio Navigation Systems

The course-glide slope system is the landing approach system for aircraft landing using instrumental radio navigation tools. The course-glide slope systems are divided into meter-range (ILS) and centimeter-range (MLS) systems.

The equipment of the ILS system consists of ground and on-board parts. The ground part of the glide path system comprises: localizer transmitter, ground and marker radio beacons. The localizer transmitter transmits at the frequencies from 108.1 to 111.9 MHz. The ILS localizer transmitter uses Morse code for identification [1].

The glide slope beacon, respectively, transmits at the frequencies from 329.3 to 335.0 MHz. The directional pattern of the localizer transmitter is horizontal, and of the glide slope beacon is vertical.

The marker beacon uses the 75 MHz radio frequency and is installed in the continuation of the landing approach flight landing strip (FLS) line axis.

The ICAO standards provide the installation of two or three radio markers. In this case, the carrier frequency is modeled.

- for a nearest marker beacon: 3000 Hz,
- for a middle marker beacon: 1300 Hz,
- for a distant marker beacon: 400 Hz. The modulation depth is 95 ± 4%.

O. I. Zavalishin et al., *Modern Requirements for Noise Immunity Aircraft Navigation Equipment*, Springer Aerospace Technology,
https://doi.org/10.1007/978-981-16-0073-9_1

The installation sites of the above-mentioned radio beacons are ≈75 m (sometimes may be absent), ≈1050 m and ≈7400 m, respectively, from FLS.

In its work the nearest radio marker transmits points, the middle radio marker transmits points and dashes, and the distant radio station transmits only dashes.

Existing examples of course-glide aircraft landing systems are: SP-50, SP-70, SP-200, etc. [2].

The listed systems form an "analog" equal-signal course-glide slope line and therefore their characteristics depend on the topography, underlying surface and on seasonality even at one airport, since the instrumental aircraft landing systems are of the category I, II and III with the decision-making height of 60 m, 30 m and 15 (0) m, respectively, and the dependence of the characteristics on the underlying surface becomes more significant with a decrease in meteorological minimum.

However, the above instrumental landing systems do not fully meet the modern flight safety requirements, in particular, at landing approach, for example, to parallel flight landing strips.

In addition, these systems are sensitive to the presence at the airfield surface of re-reflective surfaces (hangars, air terminals, administrative and technical base (ATB), air terminal, etc.).

The aircraft located on the taxiway of the airfield creates an obstacle in the arrival zone of the course-glide slope system to the landing aircraft, which makes it necessary to reduce the pass-through function of the airfield, especially in adverse weather conditions (fog, rain). All these factors adversely affect the economy and increase operating costs.

The course-glide slope system can only provide landing approach from a straight line, since the line of the equal intensity of beacons is the only one. In addition, at many airports, the existing terrain requires a more sophisticated landing approach due to the natural obstacles (mountains, forest, etc.) that cause re-reflections and shading of the radio signal.

There is also the MLS system. However, it did not find practical use because of its complexity [3, 4].

However, the above analog systems have a number of significant limitations and huge operating costs connected primarily with the need for periodic flight checks due to changes in the underlying surface. In addition, these systems must be installed with each approach course of FLS and they provide entry only on one course-glide slope line. At the same time, the LMCS installed provides 48 landing approaches from both courses of all FLS of an airfield. The positioning service at the airfield does not depend on the underlying surface, since it transmits three-dimensional digital course-glide slope lines in a single coordinate system with FLS and therefore does not require periodic flight checks. In addition, the installation of a satellite landing system may not be carried out in the FLS continuation zone, which is especially important at landing in limited areas (drill site, deck, coastal airfield, etc.).

The use of satellite aircraft approaches is the current ATC task and is supported by ICAO. However, the use of obvious advantages of satellite navigation and precision approach by GNSS should be subject to the mandatory requirements for radio systems used in ATC on the reliability of their measurements, especially at such difficult stages

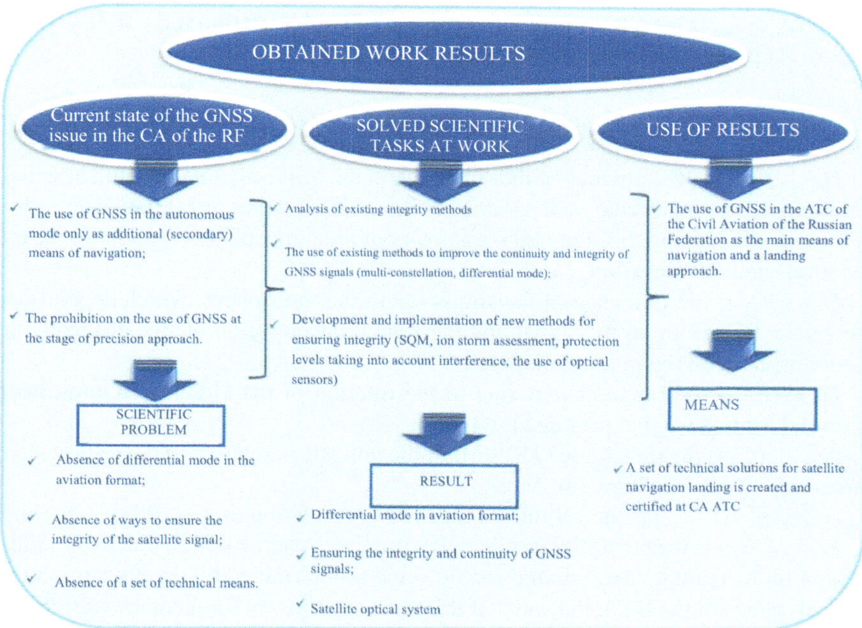

Fig. 1.1 The algorithm and the result of the work

as landing approach. To do this, it is necessary to analyze existing solutions and, if necessary, develop them to fully meet the ICAO requirements. The algorithm of the work carried out in this direction is presented in Fig. 1.1.

For the assessment of the current characteristics of satellite navigation systems, scientific studies were carried out in which a series of scientific experimental measurements were made, followed by mathematical processing of the characteristics of GNSS signals at various airport types. The measurements were carried out on the GLONASS, GPS, Galileo signals in order to assess the influence of various factors on the accuracy of determining the coordinates using the following technique described below [5].

During the use of satellite landing, the user coordinates in SRNS were determined by calculating them using the pseudo ranges (PR) to NSC.

The pseudo range (PR) measurements by the code C/A were re-calculated into units of length (multiplied by $c = 299792458$ m/s—the speed of light in vacuum).

In this case, the measured value of the PR for the nth navigation satellite (NS) and the mth base receiver (BR) can be represented as

$$D_{n,m,\text{ mes}}(t_k) = D_{n,m,\text{ true}}\left(t_k - t_{n,k,tr}\right) + D_{n,\text{ trop}}(t_k) + D_{n,\text{ ion}}(t_k) + D_{n,\text{ rot}}(t_k) + c\Delta t_{m,rec}(t_k) - c\Delta t_{n,sv}\left(t_k - t_{n,k,tr}\right) + \varepsilon_{n,m,mult}(t_k) + \varepsilon_{n,m,\text{ fuct}}(t_k), \tag{1.1}$$

where $D_{n,m,\,\mathrm{true}}\left(t_k - t_{n,k,tn}\right)$ is the true range to the nth NS from the mth BR taking into account the signal propagation time.

$$t_{n,k,tr} = t_{n,tr}(t_k) = D_{n,m,\mathrm{mes}}(t_k)/c;$$

$D_{n,\,trop}\,(t_k)$ is the influence of the signal lag in the troposphere, the value of which is determined by meteorological parameters (temperature, pressure, relative humidity of the air; when using the standard parameters of the atmosphere—height above the sea level) and NS elevation;

$D_{n,ion}(t_k)$ is the influence of the signal lag in the ionosphere, which depends on the carrier frequency of the signal and the state (concentration of free electrons) of the ionosphere on the propagation path;

$D_{n,rot}(t_k)$ is the range change due to the rotation of the Greenwich coordinate system during the signal propagation time;

$\Delta t_{m,rec}(t_k)$ is the time scale (TS) shift of the mth BR in relation to the GPS system time, $\Delta t_{m,rec} \in [-0, 5\mathrm{mc}; +0, 5\mathrm{mc}]$;

$\Delta t_{n,sv}\left(t_k - t_{n,k,tr}\right)$ is the TS shift of the nth NS in relation to the GPS system time;

$\varepsilon_{n,m,mult}(t_k)$ is the error of the code delay tracking scheme (DTS) due to the influence of multi-path; its maximum possible value is determined by the discriminatory characteristics of the DTS; the value of this error (up to 10 m for the narrowest correlator gate), its presence for measuring of PR to all visible NS and the relatively slow nature of the change over time (the correlation time up to several hundred seconds) make the multi-path error the main factor influencing the accuracy of differential corrections (DC) to PR;

$\varepsilon_{n,m,\,\mathrm{fluct}}\,(t_k)$, the fluctuation error of the DTS in nominal mode, is a Gaussian random process with zero assembly average (AA). The correlation time of the order of units in seconds and the RMSE value depends on the signal-to-noise ratio (varies with the NS elevation and the value of the wideband noise), the correlator gate width and loss in receiver paths.

The PR measurements are converted into units of length (multiplied by the wavelength $\lambda_{\mathrm{L1,gps}} = c/f_{\mathrm{L1,gps}}$, $f_{L1,\,gps} = 1575.42$ MHz for GPS or $\lambda_{L1,j,glo} = c/f_{L1,j,glo}$, $f_{L1,j,glo} = f_{L1,glo} + j \cdot \Delta f_1$, where $j = -7,\ldots,0,1,\ldots,4$ are the frequency letters (frequency channels) of the signals transmitted by the GLONASS NS in the L1 subband;

$f_{L1,\,glo} = 1602$ MHz, $\Delta f_1 = 0.5625$MHz. In this case, the measured PR value for the nth NS and the mth BR can be represented as

$$\begin{aligned}\varphi_{n,m,\,\mathrm{mes}}\,(t_k) = D_{n,m,\,\mathrm{true}}\left(t_k - t_{n,k,tr}\right) + D_{n,\,\mathrm{trop}}\,(t_k) - D_{n,\,\mathrm{ion}}\,(t_k) + D_{n,\,\mathrm{rot}}\,(t_k) \\ + c\Delta t_{m,\,\mathrm{rec}}\,(t_k) - c\Delta t_{n,sv}\left(t_k - t_{n,k,tr}\right) + \lambda_{L1} M_{n,m} + \varepsilon_{n,m,PLL}(t_k)\end{aligned} \tag{1.2}$$

where λ_{L1} is the wavelength in the L1 range (for GLONASS, NS depends on n); $M_{n,m}$ is the real PR ambiguities, which in the absence of phase jumping are constant in time;

$\varepsilon_{n,m,PLL}(t_k)$ is the tracking errors of the automatic phase synchronization scheme, including fluctuation errors and errors caused by the influence of multi-path (the total standard deviation does not usually exceed the units of centimeters).

Since GBAS uses DC to PR, the measurements of which are made with quite significant errors, to reduce the effect of DTS measurement errors, the PR values are smoothed by a linear filter taking into account the increments of the measured PP

$$D_{n,m,sm}(t_k) = \alpha D_{n,m,mes}(t_k) + (1-\alpha)\big[D_{n,m,sm}(t_{k-1}) \\ + \varphi_{n,m,mes}(t_k) - \varphi_{n,m,\text{ mes}}(t_{k-1})\big], \tag{1.3}$$

where $\alpha = \tau_e / T_{sm}$, $T_{sm} = 100\tilde{n}$ is the time constant of the smoothing filter, τ_e is the time interval between successive measurements (the duration of the epoch, which is 0.5 s for LMCS).

To analyze the solution quality of the navigation task (NT) in the autonomous mode, we used the smoothed PR (1.3) as observations with the compensation of the delay in the troposphere using the model with standard atmospheric parameters.

The NT solution was performed using the least squares method (LSM).

The vector of observations of the LSM algorithms included the smoothed PR (1.3), in which the delays in the troposphere were compensated by the model for the standard troposphere

$$D^*_{\text{trop}} = \frac{2.312}{\sin\left[\sqrt{El^2 + 0.0019}\right]}. \tag{1.4}$$

To analyze the quality of the solution of the navigation task in the autonomous mode, we used the smoothed PR with the compensation of the delay in the troposphere according to the model with the standard parameters of the atmosphere and ionosphere.

The compensation of the influence of tropospheric refraction in the differential mode was calculated as follows:

$$|\delta R_{trop \cdot p} - \delta R_{trop \cdot BS}| \le \underbrace{d\frac{1}{H\cos\varphi}(1.4588 + 0,0029611N_s) \\ - 0.3048\left[0.00586(N_s - 360)^2 + 294\right]\left(\varphi_{\text{P}}^{-2.30} - \varphi_{\text{BS}}^{-2.30}\right)}_{x}, \tag{1.5}$$

where N_s is the index of refraction, the average value of which is $N_s = 360$;

x is the parameter that determines the increase in error at small (less than 10°) elevation angles of the NSC in relation to a user.

The pseudo range increment δR_{trop} due to tropospheric delay $\delta\tau_{\text{trop}}$ is calculated as follows:

$$\begin{aligned} \delta R_{\text{trop}} &= \delta R_{\text{dry}} + \delta R_{\text{hum}} \\ \delta R_{\text{hum}} &= \frac{10^{-6}}{5} \frac{\left(-12.96 \cdot \mathrm{T} + 3.718 \cdot 10^5\right)}{\sin \sqrt{\mathrm{E}^2 + 2.25}} \frac{e}{\mathrm{T}^2} h_{\text{hum}} \\ \delta R_{dry} &= \frac{10^{-6}}{5} \frac{77.64}{\sin \sqrt{\mathrm{E}^2 + 2.25}} \frac{\mathrm{p}}{\mathrm{T}} h_{\text{dry}}, \end{aligned} \tag{1.6}$$

where E is the elevation angle of the direction UE-NSC in relation to the plane of the horizon, rad;

h_{hum} is the height of the humid layer of the troposphere;

$h_{\text{dry}} = 40{,}136 + 148.72 \cdot (T - 273.16)$ is the height of the dry layer of the troposphere, m.

Tropospheric compensation in autonomous navigation mode in the absence of meteorological data was calculated as follows:

$$\delta \tau_{\text{dry}} = \frac{2.29951}{\mathrm{c}} \mathrm{e}^{-0.000116h}; \tag{1.7}$$

$$m_{\text{cyx}}(v) = \frac{1 + \frac{a(B,D)}{1 + \frac{b(B,D)}{1 + c(B,D)}}}{\cos v + \frac{a(B,D)}{\cos v + \frac{b(B,D)}{\cos v + c(B,D)}}} + + H \left(\frac{1}{\cos \psi} - \frac{1 + \frac{a_H}{1 + \frac{b_H}{1 + c_H}}}{\cos v + \frac{a_H}{\cos v + \frac{b_H}{\cos v + c_H}}} \right); \tag{1.8}$$

$$m_{Bx}(v) = \frac{1 + \frac{a_{\text{ВП}}}{1 + \frac{b^{m}_{\text{ВП}}}{1 + \frac{\mu}{c_{\text{ВП}}}}}}{\cos v + \frac{a_{\text{ВП}}}{\cos v + c_{\text{ВП}}}} \tag{1.9}$$

$$\begin{aligned} a(B, D) &= \hat{a} - ap \cos\left(2\pi \frac{D - D_0}{365, 25}\right); \\ b\left(B, D_1'\right. &= \hat{b} - b_p \cos\left(2\pi \frac{D - D_0}{365, 25}\right); \\ c(B, D) &= \hat{c} - c_p \cos\left(2\pi \frac{D - D_0}{365, 25}\right), \end{aligned} \tag{1.10}$$

where $m_{dry}(\upsilon)$, $m_{hum}(\upsilon)$ are the mapping functions for the dry and humid components of the vertical tropospheric delay;

$\hat{a}$, $\hat{b}$, $\hat{c}$ are the average coefficients of the mapping function for the dry component;

a_p, b_p, c_p are the amplitude coefficients of the mapping function of the dry component;

$a_н = 2.53 \cdot 10^{-5}$, $b_н = 5.49 \cdot 10^{-3}$, $c_н = 1.14 \cdot 10^{-3}$ are the altitude coefficients of the mapping function for the dry component;

a_{hum}, b_{hum}, c_{hum} are the coefficients of the mapping function of the humid component; D_0 is the starting day of the year.

Ionospheric refraction in the autonomous navigation mode is calculated using the model in which the ionosphere is considered as a material point located at a height $h_{IT} = 350$ km

$$\delta R_{\text{ion}} = c\delta\tau_{\text{ion}} = \frac{1}{\cos\psi}\frac{40,3}{f^2}N_\Sigma;\ \sin\psi = \frac{R_3}{R_3 + h_{IT}}\sin\varphi. \quad (1.11)$$

Here ψ is the zenith angle in the subionospheric point;
N_Σ is the vertical density of the electron content;
R_3 is the radius of the Earth;
φ is the Zenith angle at the point of location of the user.

The results of the experiments on the estimation of errors in determination of the coordinates in the autonomous mode from the 4.12.2016 to 21.12.2016 records are presented for convenience of perception in a graphical form in Figs. 1.2a, b, c and 1.3a, b, c, respectively.

For completeness of analysis in all expected operating conditions, the following was done for the experiment [6]:

- Three airports with different operating conditions were selected:
  (A) The mountain airport ("Gorno-Altaisk");
  (B) The subpolar airport (Antarctica Icefield "Novolazarevskaya");
  (C) The plain airport of the city of Samara ("Kurumoch");
- The 100-Hz 316-channel satellite navigation receivers GPS, GLONASS, Galileo and ChockRing (RingAnt) antennas with protection against reflections were installed at the points with known geodetic coordinates (PZ 90.11). To assess the multi-path propagation of GNSS signals at each airport, two satellite receivers were installed (each at its own antenna).

The coordinates were measured in the PZ-90 system (GLONASS).
The coordinate system for GPS was re-calculated using the matrix:

$$\begin{bmatrix} X \\ Y \\ Z \end{bmatrix}_{\text{WGS-84}} = \begin{bmatrix} -0,03 \\ -0,27 \\ -0,92 \end{bmatrix} + \left(1 + 0,10 \times 10^{-6}\right)\begin{bmatrix} 1 & -0,07 \times 10^{-6} & 0 \\ +0,07 \times 10^{-6} & 1 & 0 \\ 0 & 0 & 1 \end{bmatrix}\begin{bmatrix} X \\ Y \\ Z \end{bmatrix}_{\text{PZ-90.02}} \quad (1.12)$$

The records were made at an annual time interval, which allowed to estimate the seasonal fluctuations due to changes in temperature, daily fluctuations due to changes in satellite geometry and ionospheric gradient (sunrise, sunset), as well as the effect of ephemeris changes.

The qualitative analysis of the error data shows that in the autonomous mode the accuracy of the GLONASS signals is significantly worse than that of the GPS signals, especially from the results of processing of records dated 21.12.2016.

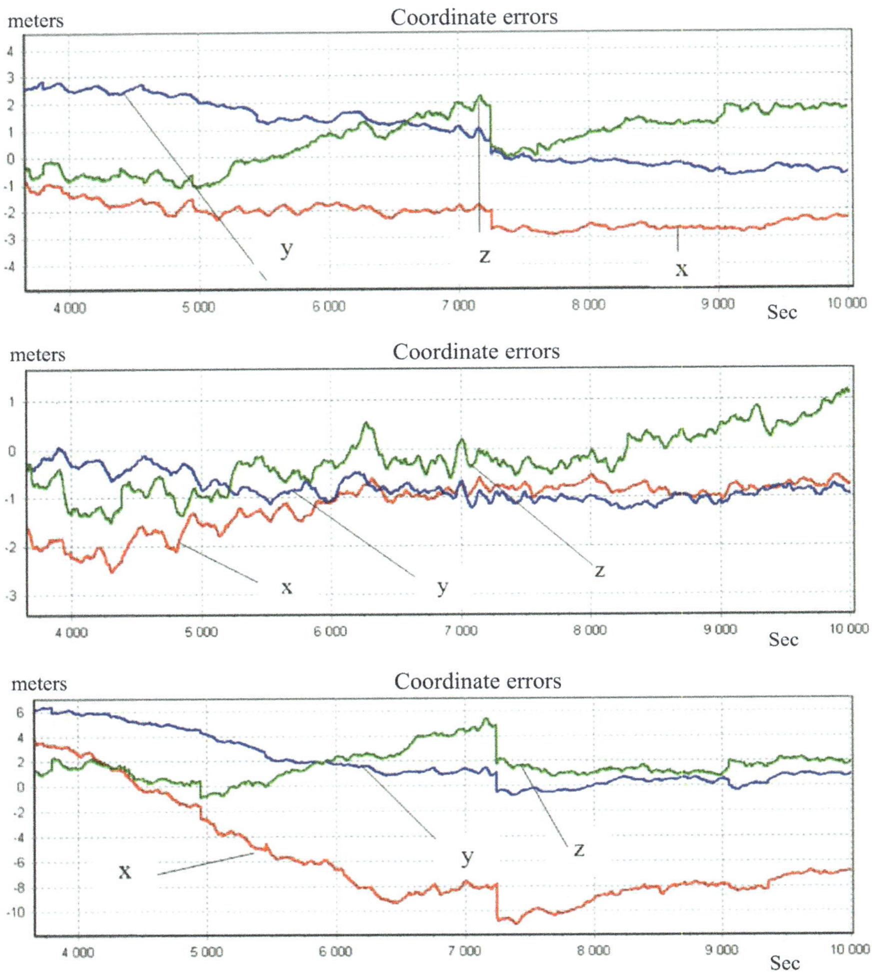

Fig. 1.2 **a** The errors of determination of the coordinates in the autonomous mode GPS + GLONASS according to measurements made in the laboratory of RDPF Spektr (Moscow) on 14.12.2016 **b** The errors of determination of the coordinates in the autonomous mode GPS according to measurements made in the laboratory of RDPF Spektr (Moscow) on 14.12.2016 **c** The errors of determination of the coordinates in the autonomous mode GLONASS according to measurements made in the laboratory of RDPF Spektr (Moscow) on 14.12.2016.

At joint use of signals from both systems, the quality of the navigation task solution in the autonomous mode is also worse in comparison with GPS.

The presented implementations cannot be considered stationary random processes. Consequently, the statistical characteristics formally obtained as a result of processing of these implementations contain significant methodological errors

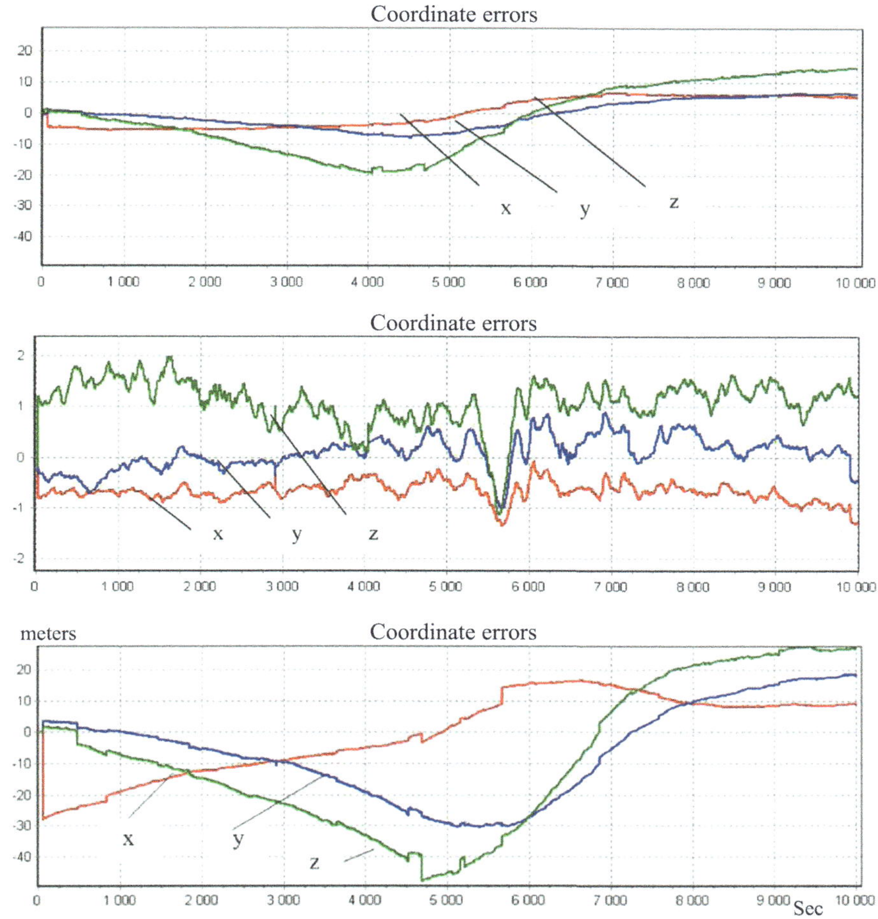

Fig. 1.3 **a** The errors of determination of the coordinates in the autonomous mode GPS + GLONASS according to measurements made in the laboratory of RDPF Spektr (Moscow) on 21.12.2016 **b** The errors of determination of the coordinates in the autonomous mode GPS according to measurements made in the laboratory of RDPF Spektr (Moscow) on 21.12.2016 **c** The errors of determination of the coordinates in the autonomous mode GLONASS according to measurements made in the laboratory of RDPF Spektr (Moscow) on 21.12.2016.

and can only serve as approximate indicators of the quality of the navigation task solution. The records of 14.12.2016 and 21.12.2016 are given in Tables 1.1 and 1.2.

As follows from the analysis of the statistical characteristics of the coordinates determination errors, the RMSD of errors of determination of the coordinates using the GLONASS signals are several times greater than the RMSD of errors of determination of the coordinates by GPS (3–8 times for the records from 14.12.2016 and 50 times for the records from 21.12.2016) [3, 7–9].

Table 1.1 The statistical characteristics obtained as a result of the processing of navigation measurements made in the laboratory of RDPF Spektr (Moscow) on 14.12.2016

Operation mode		GPS + GLONASS	GPS	GLONASS
X coordinate	AA, m	−2.17	−1.16	−5.89
	RMSD, m	0.48	0.47	3.97
Y coordinate	AA, m	0.86	−0.81	1.89
	RMSD, m	1.16	0.29	2.09
Z coordinate	AA, m	0.59	−0.24	1.59
	RMSD, m	0.94	0.57	1.19

Table. 1.2 The statistical characteristics obtained as a result of the processing of navigation measurements made in the laboratory of RDPF Spektr (Moscow) on 21.12.2016

Operation mode		GPS + GLONASS	GPS	GLONASS
X coordinate	AA, m	0.72	−0.66	1.94
	RMSD, m	4.88	0.20	10.70
Y coordinate	AA, m	−0.54	0.14	−6.9
	RMSD, m	4.76	0.29	15.50
Z coordinate	AA, m	−1.80	1.05	−9.06
	RMSD, m	11.57	0.45	24.43

The statistical characteristics were calculated according to the obtained measurements:

$$\mathrm{m_n} = 1/\mathrm{n} * \sum_{i=1}^{N} * \varepsilon_\mathrm{i}$$

Since the error in determination of the coordinates in the autonomous mode on GPS signals does not exceed the units in meters, it can be said that in this case the observed signals have good ephemerical-time support quality (ETS), the delay in the troposphere is compensated on the model with acceptable accuracy and the ionospheric delay is insignificant and does not lead to catastrophic consequences in the navigation task solution. Therefore, it can be assumed that the main source of errors in determination of the coordinates on the GLONASS signals is ETS errors [10].

At the same time, the obtained experimental data testify to the inadmissibility of landing operations on GNSS (RMSD is 11.57 m).

The well-known effective method for improving the accuracy of determination of the coordinates is the differential mode method. However, the GNSS differential mode itself does not guarantee integrity at provision of the accuracy required by ICAO.

To ensure the integrity of the differential mode, it is necessary to estimate the error of the differential corrections to PR for each range-measuring source (satellite). In order to analyze the dynamics of changes in the statistical characteristics of differential corrections to PR, a series of experiments was carried out using a highly accurate geodesic reference point.

The intermediate values of DC for the nth NS and the mth BR without the compensation of the BR TS shift are calculated by the formulas

$$\Delta'_{n,m} = D_{n,m,calc}(t_k) - \left[D_{n,m,sm}(t_k) - D^*_{n,m,rot}(t_k) + c\Delta t^*_{n,sv}\left(t_k - t_{n,k,tr}\right)\right]. \tag{1.13}$$

To compensate the decrease in the BR TS, the arithmetic average of corrections for all visible NS is often used

$$\Delta t^*_{m,rec}(t_k) = -\frac{1}{\tilde{n}N(t_k)} \sum_{i \in S_e} \Delta'_{i,m}(t_k), \tag{1.14}$$

where S_c is the set of range-measuring sources (NS signals) that are not rejected by the integrity control algorithms, and $N(t_k)$ is the current number of elements of this set.

The correction transmitted to the user (PRC) to reduce the effect of multi-path is formed as [4, 11]

$$\Delta_n(t_k) = \frac{1}{M_n(t_k)} \sum_{m \in S_n} \Delta_{n,m}(t_k), \tag{1.15}$$

where S_n is the set of BR used for the formation of the nth DC, and $M_n \geq 2$ is the number of elements of this set.

Figure 1.4 shows the number of visible range-measuring sources by the GNSS base receiver of the base station of the GBAS local functional supplement for landing operations on the I–III categories of ICAO and terminal (APV, NPA) operations on satellite signals of GLONASS/GPS/Galileo (L_1).

Figure 1.5 presents the results of the statistical processing of experimental data of the differential mode of the GBAS equipment for performing a precision landing approach according to the I–III categories of ICAO and terminal operations at the airfield using GNSS GLONASS/GPS/Galileo satellite signals (a/p "Emelyanovo").

The errors in determination of the coordinates in the differential mode, obtained on the basis of experimental measurements over a long time interval, including during the flight tests at the Ramenskoye and Emelyanovo airfields on the ACLs Il-76 and An-24 are presented in Fig. 1.6.

Figure 1.6 presents the results of the statistical processing of empirical data of the coordinates determination errors on the GBAS data in the differential mode.

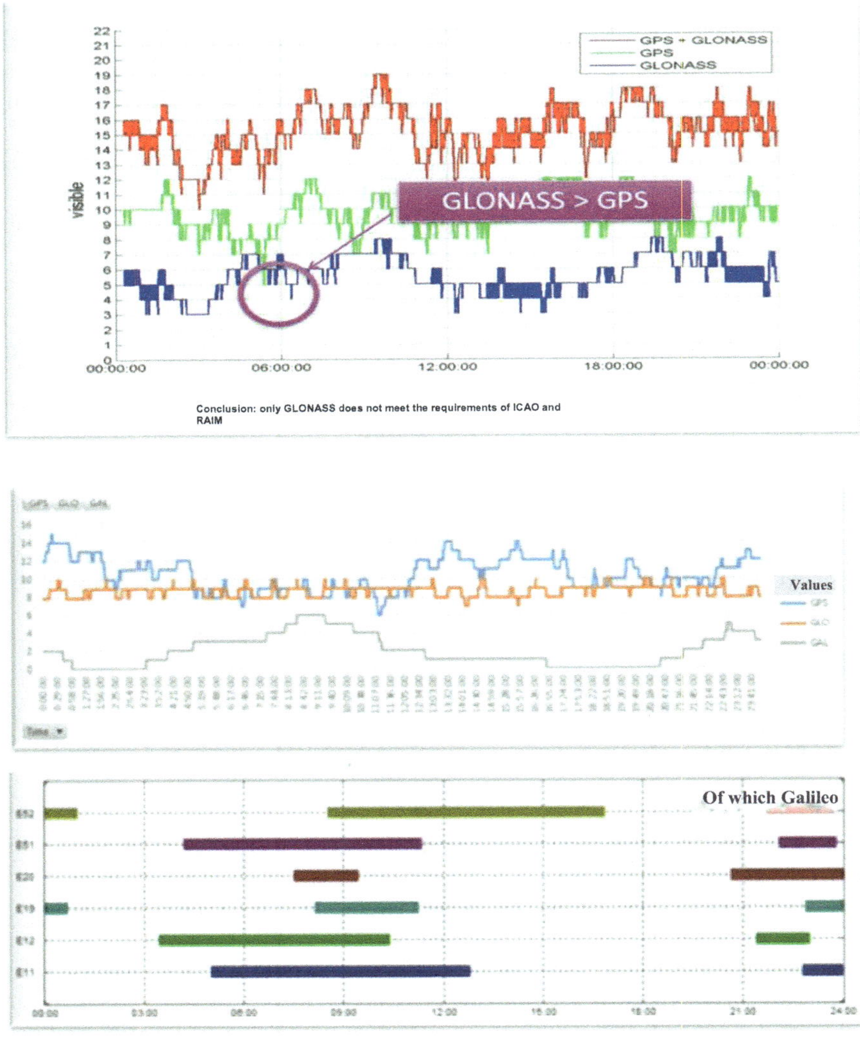

Fig. 1.4 Daily visibility of GNSS satellites at the Yemelyanovo and Ramenskoye airports by GBAS

As it follows from the analysis of the statistical characteristics of the errors of determination of the coordinates, the RMSD of the errors of determination of the coordinates on the GLONASS signals is several times higher than the RMSD of the errors of determination of the coordinates of the GPS system (3–8 times). Accuracy of the differential mode (RMSD) is by 10 times higher than the accuracy of the autonomous mode, amounting to RMSD ≤ 0.5 m, which meets the requirements of ICAO (Annex 10, Volume I "Aviation Telecommunications").

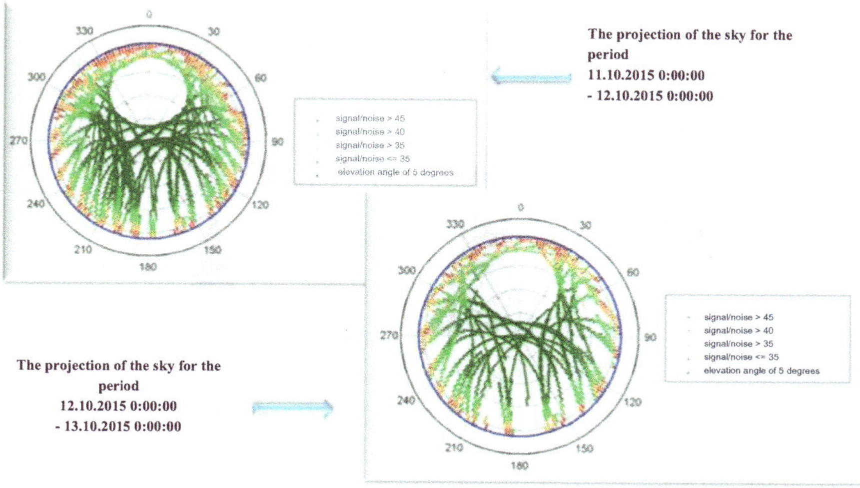

Fig. 1.5 The results of statistical processing of GBAS data to ensure the landing approach on the I–III categories of ICAO

In civil aviation much attention is paid to the possibility of obtaining benefits from the use of high-precision modern GLONASS/GPS navigation satellite systems at all stages of flight, including precision landing approach [12–14].

However, the weak power of the satellite signal (by the criterion of mass and power consumption of the transmitter on the satellite) makes the signal dependent on external factors (disturbance of the ionosphere, state of the troposphere, solar flares, magnetic storms and industrial radio interference). Therefore, the important task at the use of GNSS signals is to ensure the integrity of satellite navigation signals.

The integrity assurance measure is the probability of detection of a failure within a specified time and informing a user on it. ICAO standards set high integrity requirements (confidence-building measures) for GNSS signals and their quality (SQM).

The results of the experiments presented in Table 1.1 showed that the autonomous satellite navigation field of GLONASS/GPS does not meet the ICAO integrity requirements at *all* stages of the aircraft flight. The necessary characteristics of the accuracy and integrity of the signals are provided only with the joint use of space groups and functional augmentations to the GNSS.

The current status of the issue and the proposed solution to ensure the characteristics of GNSS signals when performing satellite landing approaches as the main means are presented in Fig. 1.7.

The experimental evaluation of radio signal delay factors based on the analysis of a posteriori pseudo range measurements in various regions of the Russian Federation allowed us to give the qualitative assessment of the sources of errors and ways of compensation of each of the errors (Figs. 1.8, 1.9, 1.10, 1.11 and 1.12).

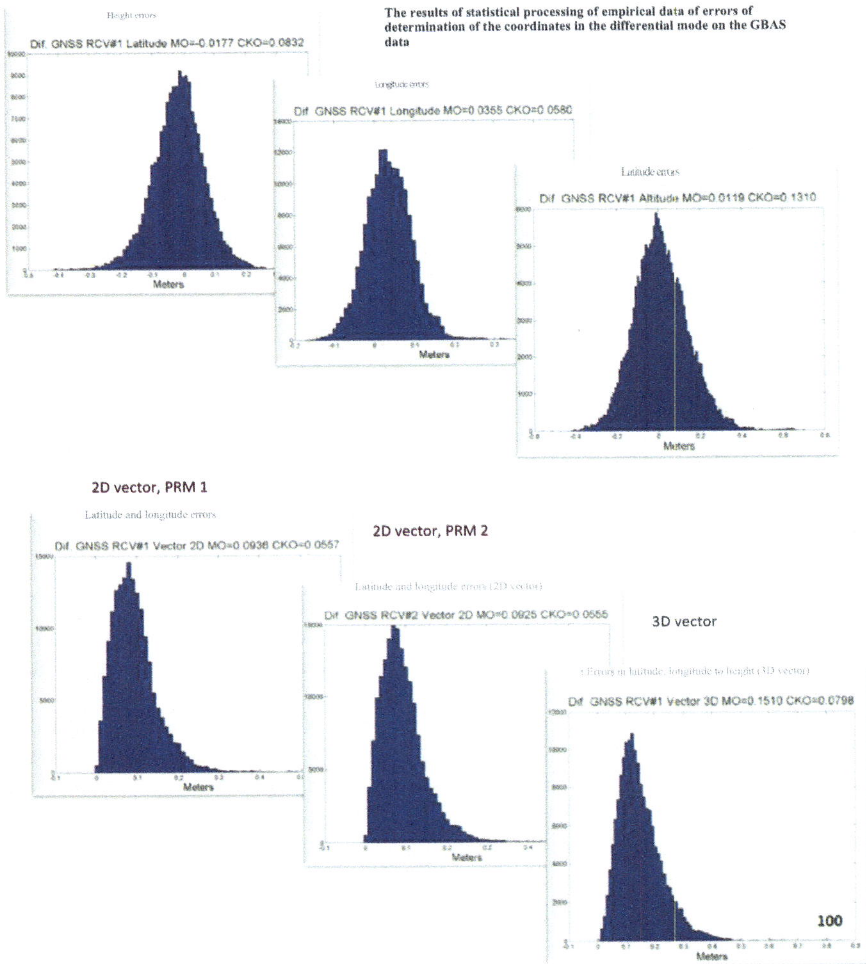

Fig. 1.6 The results of processing the data of errors in determination of the coordinates according to the GBAS data in 2D and 3D

One of the effective ways to improve the integrity of GNSS signals is the differential mode of GNSS operation, since in this mode the main PR measurement errors (ephemerical errors, satellite clock errors, tropospheric and ionospheric delays) are almost the same for the ground station and the user.

Since the ground station is stationary at the point with known coordinates, it is quite simple to implement the algorithms for monitoring of anomalous distortions of PR and/or PP, thereby ensuring the required integrity characteristics.

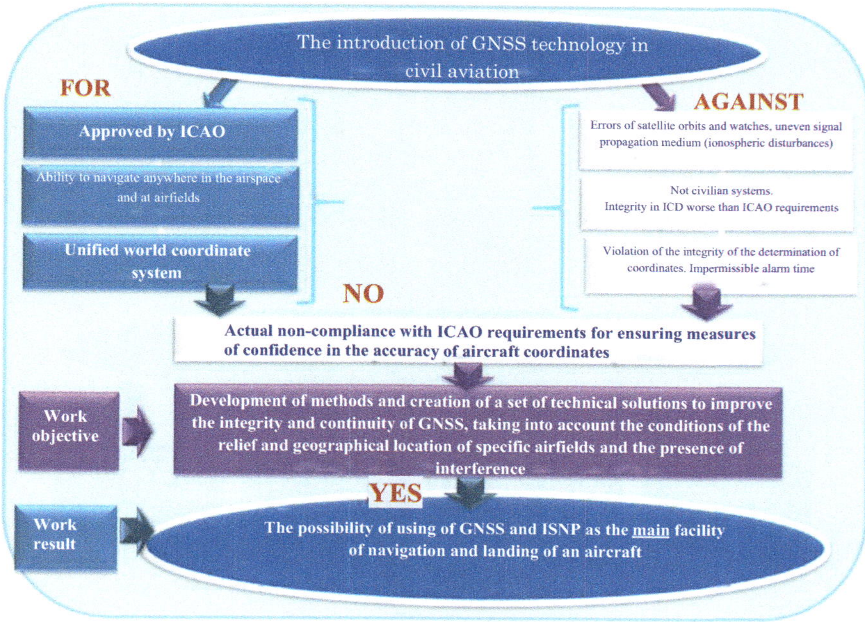

Fig. 1.7 The use of GNSS as the primary means of navigation and landing

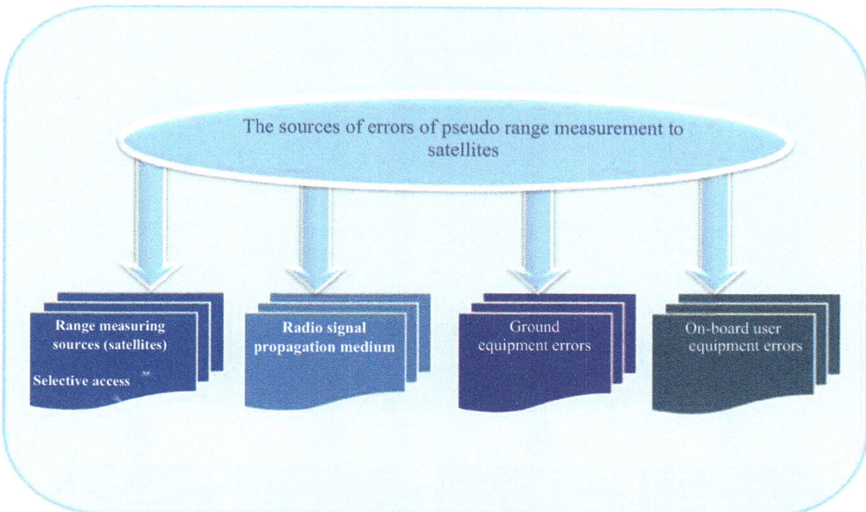

Fig. 1.8 The sources of errors of pseudo range measurement to satellites

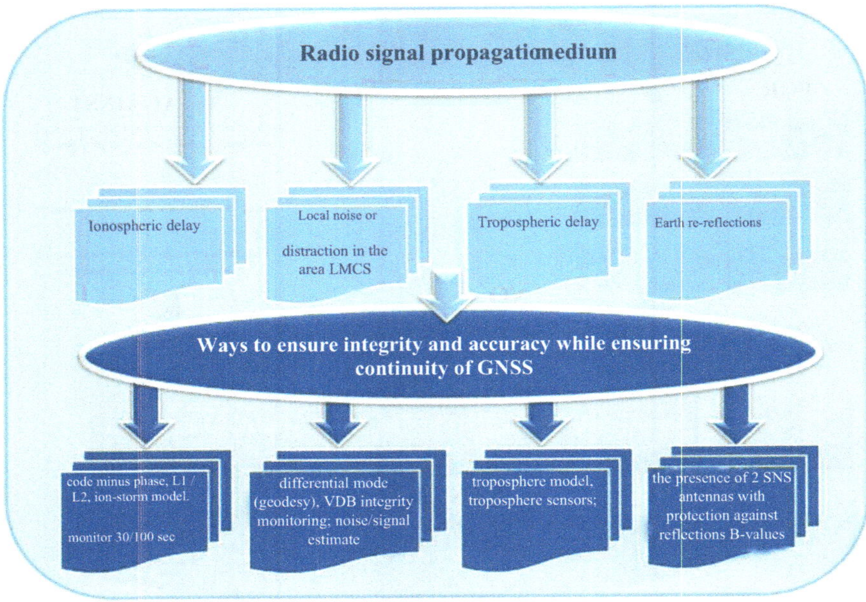

Fig. 1.9 The radio signal propagation environment and ways to ensure the integrity and accuracy of the obtained coordinates

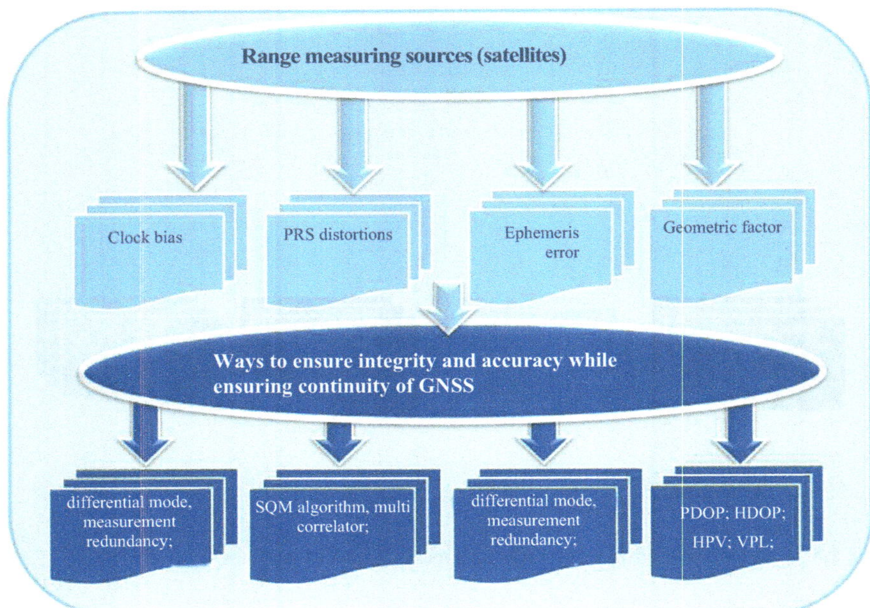

Fig. 1.10 The range-measuring sources (satellites) and ways to ensure the integrity and accuracy of the obtained coordinates

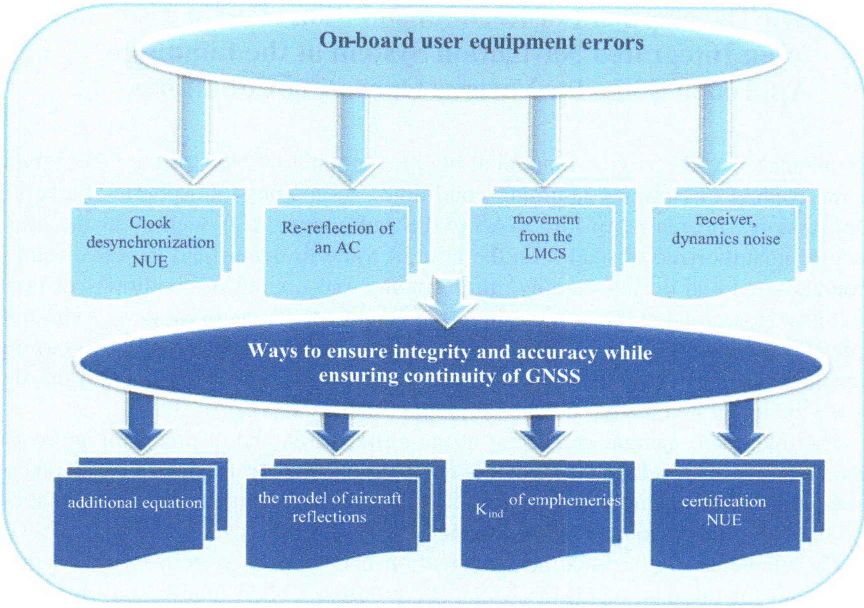

Fig. 1.11 The errors of on-board user equipment and ways to ensure the integrity and accuracy of the obtained coordinates

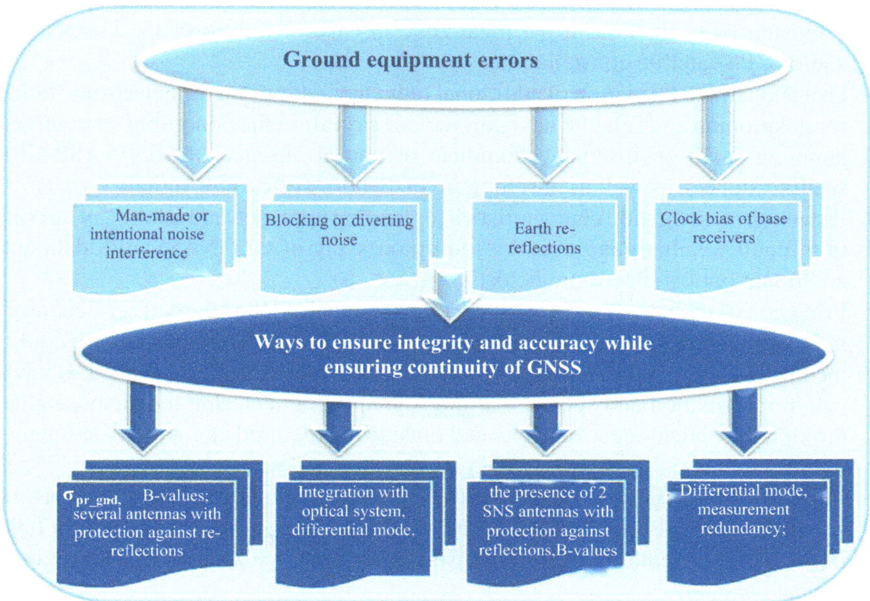

Fig. 1.12 The errors of ground equipment and ways to ensure the integrity and accuracy of the obtained coordinates

1.2 The Development of Recommendations for the Use of an Integrated Navigation System at the Landing Approach Stage for Various Operating Conditions

To enhance the integrity of functional augmentations to GNSS of the GBAS-type in the context of complex jamming conditions, anomalous ionospheric "flickers", suppression/ "decrease" of GLONASS/GPS radio signals, as well as in the presence of unauthorized obstacles on the aircraft route in the airfield area (especially mountainous) and during landing, after landing and taxiing in conditions of poor visibility (fog/clouded sky) and at night, as well as for transmission via radio and optical channels to an aircraft board of navigation information and special warning signals about impermissible deviations from the line of a given path, it is proposed to use the integrated navigation and landing system (INLS).

The integrated navigation system at the airfield area and at precision approach (INLS) has been developed for satellite navigation and aircraft landing based on co-processing of data from GNSS sources, LMCS land-based functional augmentation and optical multi-band navigation channel.

The elements of the integrated navigation and landing system based on dual processing of the data of GLONASS/GPS systems, LMCS land-based functional augmentation and optical multi-band navigation channel (INLS) are used in landing approaching in severe weather conditions (fog, cloudiness) for the following tasks [15]:

- Provision of navigation in the local zone in case of failure of the GLONASS and/or GPS satellite constellation;
- Provision of the formation of additional radio navigation and optoelectronic fields for positioning and reliable navigation of an aircraft in the conditions of complex jamming radio environment, complete or partial absence of GLONASS/GPS satellite signals, as well as at setting up of "leading off" radio signals;
- Provision of navigation characteristics for landing approach and landing of aircraft in difficult weather conditions by joint processing of GLONASS/GPS data and additional optical-electronic fields;
- Provision of transmission through the radio channels to the board of service information, including differential corrections for GPS/GLONASS navigation equipment, FAS landing data set in accordance with ICAO standards, as well as navigation information and special warning signals about entering the airspace zone through the optical-electronic channel outside the required navigation parameters (RNP);
- Provision of navigation in emergency situations, including in the conditions of suppression of GPS/GLONASS signals by including an optical navigation field in the radio navigation field and transmission of service information through an optical channel;
- Positioning of aircraft on the FLS approach in a radio-independent way;

- Delivery of information for sound and light signaling in case of obstacles in the course of a flight at an unsafe distance, detected by the use of ultraviolet (UV) and infrared (IR) receivers;
- Continuous automatic monitoring of the system;
- Land-based complex processing of navigation information from GLONASS/GPS and optoelectronic sensors and delivery of its results to the air traffic management (ATM) services of the airfield;
- Transfer of coordinate information from optoelectronic sensors to an on-board computer for complex processing of GLONASS/GPS data on the ARINC 429 protocol.

The developed equipment of the integrated navigation and landing system based on the joint processing of GLONASS/GPS data, the GBAS land-based augmentation system and an optical multi-band navigation channel can be used as radio engineering and signaling equipment for CA airfields, special, test and state aviation, heliports and helicopter sites, as well as in airlines and aircraft/helicopter construction design bureaus [16].

The users and objects of application of this equipment are airlines, air traffic control services (ATC), FSUE "State ATM Corporation", airfield services, technical services of helipads, heliports, rapidly deployed helipads, especially in the northern and far-eastern regions with severe weather conditions, low cloudiness, frequent fogs and difficult ground profile.

The requirements for the quantitative characteristics of INLS navigation tools comply with ICAO standards and are reflected in Table 1.3.

The INLS deployment scheme is presented in Fig. 1.13.

To ensure the specified accuracy, the sources of navigation sensors in the minimum configuration should be located as follows:

In the conditions of suppression of satellite navigation signals, the provision of land-based optical markers of the optoelectronic system (OES) (from three to six pieces) is necessary to ensure navigation and landing functions [17, 18].

The OES is able to determine the location of the aircraft, both on the ground and in the air, as well as to determine obstacles (electrical wires, pipes, cars on FLS, etc.), including in conditions of cloudiness and fog, especially in adverse weather conditions along the route of the aircraft. In international practice this technology is called enhanced vision system (EVS).

The implementation of this technology is important for high-precision navigation, visualization of obstacles, ensuring landing and also makes it possible to eliminate the lag of the Russian industry in the development of EVS and SNS technologies for civil, state and business aviation.

To provide navigation when the aircraft is in the airfield zone and on landing after the entry point to the glide path on the final approach track at a distance of up to 5000 m from FLS, the optoelectronic subsystem of the ultraviolet (UV) range is offered as an additional channel to the satellite channel (and integrated with it), which provides additional control of the satellite radio channel in a normal situation and navigation in case of radio channel failure [9, 19].

Table 1.3 Requirements for the quantitative characteristics of the navigation equipment of the INLS

Typical operation [77]	Horizontal accuracy 95%	Vertical accuracy 95%	Integrity	Time to warning (s)	Continuity	Operational readiness	Relevant RNP type (nm)
En-route	3700 m (2 nm)	Not assigned	0.9999999 for 60 min	300	From 0.9999 to 0.99999999 per hour	From 0.99 to 0.99999	From 10 to 20
En-route near the airfield	740 m (0.4 nm)	Not assigned	0.9999999 for 60 min	15	From 0.9999 to 0.99999999 per hour	From 0.99 to 0.99999	From 1 to 5
Initial approach [2]. Intermediate approach. Non-precision approach (NPA) Taking off	220 m (720 ft.)	Not assigned	0.9999999 for 60 min	10	From 0.9999 to 0.99999999 per hour	From 0.99 to 0.99999	From 0.5 to 0.3
Non-precision landing approach procedures with vertical guidance (APV-I)	220 m (720 ft.)	20 m (66 ft.)	0.9999998 per approach	10	0,999,992 in any 15 s	From 0.99 to 0.99999	0.3/125
Non-precision landing approach procedures with vertical guidance (APV-II)	16 m (52 ft)	8.0 m (26 ft.)	0.9999998 per approach	6	0.999992 in every 15 s	From 0.99 to 0.99999	0.03/50
Categorized precision landing approach on the category I	16 m (52 ft)	6.0 m to 4.0 m (20–13 ft)	0.9999998 per approach	6	0.999992 in every 15 s	From 0.99 to 0.99999	0.02/40

The system is based on two typical modules: UV pulse emitter based on an excimer lamp (UV beacon) and a UV receiver based on monophoton technologies.

In the minimum configuration, one wide-angle photodetector synchronized with one UV widely directed source is placed on-board. The receiver performs the functions of a controller and exchanges with on-board equipment.

The algorithm of the system consists of four modes that can be used separately or together:

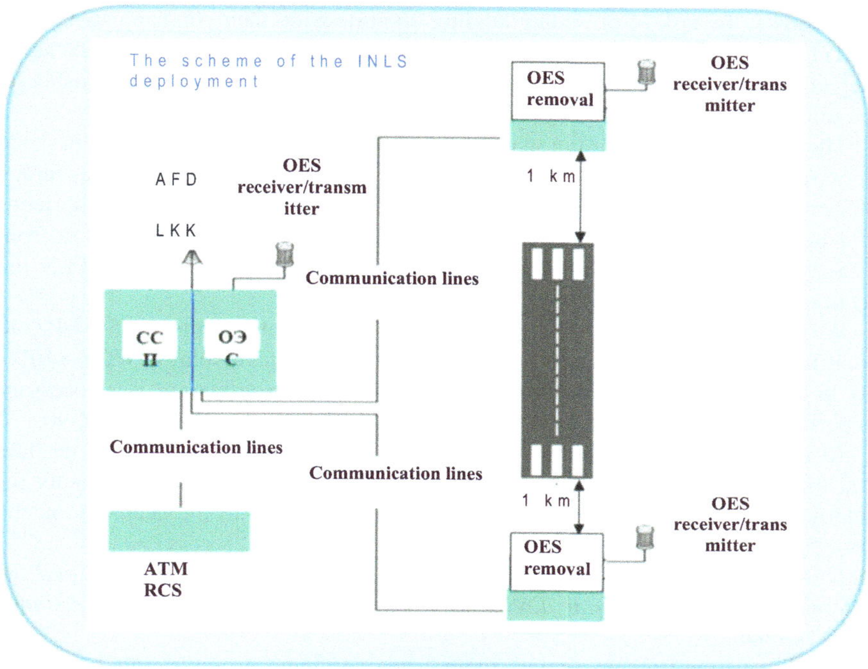

Fig. 1.13 The scheme of the INLS deployment OES—optoelectronic system, SLS—satellite landing system

(A) The observation by land-based UV receivers of beacons at an aircraft: This ensures the detection of an aircraft and the determination of its position in space;
(B) The observation by the UV receiver of an aircraft of the beacons at the airfield and the determination of the angular directions to the beacons and the angular distance between the beacons;
(C) The exchange of data obtained by the ground and on-board subsystems between them, and the formation of integrated information;
(D) Synchronous measurements of the on-board and ground subsystems and the joint measurement with high accuracy of the ranges between the aircraft and the ground beacons.

It should be noted that the measurements are carried out in their own coordinate systems, and it is necessary to apply carefully the transformations to other systems so as not to impair the accuracy of measurements during transformations.

Observation by land-based UV receivers of beacons at an aircraft.

Receivers should be placed with the widest possible base on the sides from the FLS and their view axes should be directed toward the landing approach glide path.

If you set the task of providing landing on glide path, then you can use narrow-angle receivers, for example, with a field of sight of 12°. They will provide the accuracy of the angular measurement not worse than 2×10^{-4} rad (in normal atmosphere, the refraction on clouds and fogs is a subject of specialized consideration).

The position of an aircraft in space will be determined by the intersection in space of two cones with the disclosure of 2×10^{-4} rad each. This will provide the accuracy of a single measurement in directions perpendicular to the FLS (height and horizon) of 1 m of the standard deviation when an aircraft is at a distance of 5000 m from FLS. When approaching the decision point in the area of 1000 m from the FLS, the accuracy increases.

We note that the locations of the receivers can be tabulated in the WGS-84/PZ-90 coordinate system. This is the basis for an aircraft coordinates display in WGS-84/PZ-90. In this case, for the transformation of the angular measurements, it is necessary to perform transformations to the aircraft coordinate system in WGS-84/PZ-90.

At the same time, the correctness of the conversion should consist of the fact that the worst accuracy of determination of the goniometrical measurements for determination of the range should not lead to impairment in the accuracy on the horizon and height in the WGS-84/PZ-90 system [20].

Observation by the aircraft UV receiver of the beacons at the airfield and the determination of the angular directions to the beacons and the angular distance between them.

On observation from an aircraft, there is no coordinate system rigidly connected with the earth, and the observation is carried out in the coordinate system connected with an aircraft construction line.

To establish a balance between the accuracy of determination of the angular coordinates, the need to ensure continuous sighting of the minimum configuration of the UV beacons and the installation of the basis among UV beacons for a single aerodrome for an on-board UV detector, the optical system can be selected with a field of view of 60° or 30°. The accuracy of the measurement of the angular coordinates of the beacons at the FLS will be 0.005° and 0.01°, respectively.

These measurements can be presented in several interpretations:

- The measurement of the direction to the beginning of the FLS for the landing stage on the glide path, or the direction to the airfield (some reference point on the airfield) moving in a circle (if the aircraft is in the field of view of the receiver);
- The measurement of the relative angle among the beacons, which, while moving along the glide path, actually allows to calculate the range at the beginning of the FLS.

  For a range of 3000 m, the error in this case will be not more than 3 m, while for the range of decision making of 1000 m, the error will be about 1 m.
- The measurement of the position of the plane of an aircraft in relation to the plane of the FLS by the position in the angular space of the line between the beacons on the FLS.

Its accuracy will be approximately 10.3 rad, in the range of 3000 m and will reach 0.3 mrad in the decision making range of 1000 m. This is actually a direct measurement of the aircraft roll in relation to the FLS plane [21].

The position of the image of the third beacon at the FLS with respect to the image of the line between the first two beacons can directly determine the tangage of an aircraft in relation to the surface of the FLS. The accuracy of this measurement due to the close angular position of the vertical image of the beacons at landing on the glide path will be significantly less. Thus, the measurements from the board show good instantaneous (not calculated by the prolongation of the trajectory) coordinate values of the construction line of an aircraft in relation to a FLS.

In normal operation mode, the operation of the INLS equipment is carried out automatically and does not require the intervention of operators. Control over the information generated by the ground INLS subsystem is carried out by the operators of ATS by visual information.

The general structure of work on the use of INLS is shown in Fig. 1.14.

The functioning of the INLS in the conditions of the complete absence of the GLONASS system signals can be considered in three variants:

The first variant is intentional or unintentional turning off of space transmitters;

The second variant is radio interference in the area of aircraft operations at GLONASS frequencies (failure of space and land-based distance sources);

The third variant is the deliberate transmission of false navigation signals.

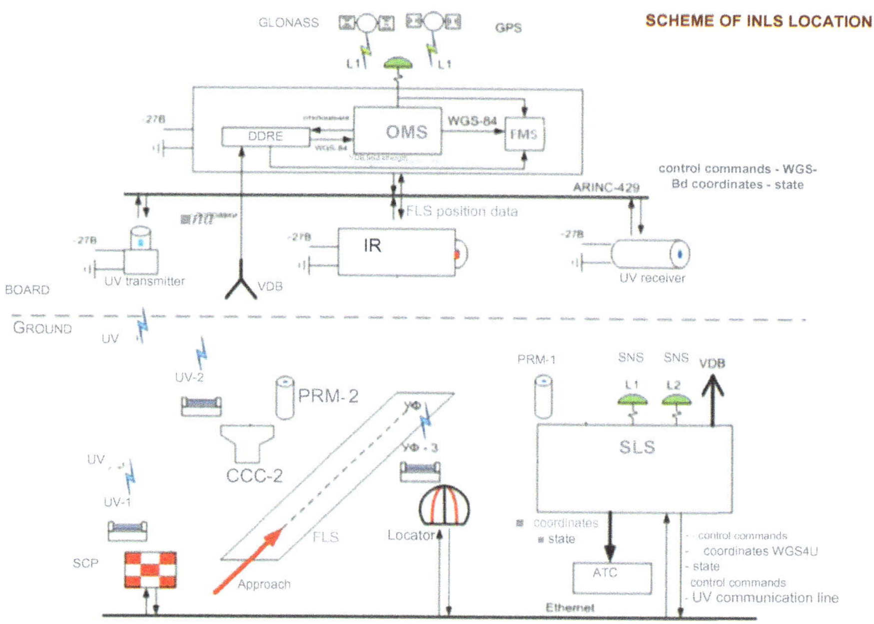

Fig. 1.14 The general scheme of the deployment of the INLS equipment at Ramenskoye a/f

In the first variant the signals of three optoelectronic modulated transceivers located on the ground are used for navigation determination.

In the second variant, at an unacceptable level of interference for range-measurement navigation spacecraft (NSC), the navigation determination on-board the aircraft is carried out by measuring the ranges of up to three separated land-based modulated optoelectronic transceivers.

In the third variant, the deliberate distortion of the navigation information will be detected by the LMCS station (due to its geodetic connection and integrity control algorithms), and information about this event will be transmitted to an aircraft board and the ATM service. The navigation determination in this case will be carried out on the optoelectronic subsystem without a significant reduction in accuracy characteristics.

The use of an integrated system will provide navigation and landing in the conditions of all three failure scenarios of the radio navigation subsystem.

The fulfilment of these functions was confirmed during flight tests at FSUE Gromov Flight Research Institute and INLS modeling [14, 16].

The general scheme of the deployment of ground and on-board INLS equipment is shown in Fig. 1.14.

It is important to note that the emission of UV beacons occurs in the form of modulated signals, which allow determination of an aircraft location simultaneously to carry out two-way data transmission, and in the solar-blind range, which is not subject to “flashing”.

Integration of information about an aircraft location from two sources of different physical nature (radio signals and optical signals) significantly increases the integrity, continuity and, as a result, availability of satellite navigation systems and precision landing approach along with maintaining the accuracy requirements.

With this method of implementing of an integrated system additional functions appear in addition to landing [22]:

- data transmission line of the UV range (communication),
- observation function, and
- navigation.

1.3 The Methods for Improving the Noise Immunity of the Integrated Navigation System in the Airfield Area Using Precision Approach Based on the Optimal Placement of Ground Equipment and Three-Dimensional Visualization of the Ground Profile On-Board an Aircraft

For a minimal configuration of facilities, we can offer the following construction. At the airfield there are two receivers with the field of view of 30° each and axes directed along the glide path. The base between them is about 2500 m. Three UV sources are also located. Two of them on the side surfaces of a FLS next to the receivers, and one source at the beginning of a FLS. At the request 1–2 beacons are placed at the glide path route outside the airport.

One receiver with the field of view of 60° and one wide directional UV emitter are placed on-board. The UV communication and synchronization are established between the board and the ground to measure the ranges between an aircraft and the ground beacons. As a result, it is possible to determine the position of the center of mass of an aircraft in the WGS-84 with the accuracy better than 1 m at the range of 3000 m and to obtain direct information on the COURSE, ROLL and TANGAGE without the use of trajectory analysis. Additionally, collision prevention features are provided.

In the exchange protocol with the on-board radio equipment, it is advisable to provide information fields for the transmission of these data, possibly in raw form, to minimize the loss of accuracy on conversions and processing [23].

The general scheme of the functioning of the system with the best construction is shown in Figs. 1.15 and 1.16.

The increase in accuracy of measurement is achieved by installation of a larger number of beacons. For example, installation of additional beacons on the glide path makes the task of determination of height and range correct and significantly increases the accuracy of the goniometric measurements from an aircraft board. Taking into account the requirement of continuous provision of sighting of at least three UV beacons to ensure navigation, as well as the need to ensure the possibility of a two-way approach to FLS according to the geometric analysis, it is necessary to install at least 8 posts with UV emitters (beacons) (Fig. 1.17).

At the same time, UV posts should be grouped into at least two triangular contours with different bases: sighting of the primary contour with the base of about 1000 m (if the UV detector with the field of view of 60° is used on an aircraft board) provides navigation in the "far" zone—more than 1500 m from the edge of a FLS, while the second contour with half the base provides the navigation in the "near" zone—less than 1500 m from the edge of a FLS, as well as taxiing on FLS [24].

Since each UV post consists of two UV beacons, whose radiation patterns are deployed in different directions (bidirectional radiation), the contours of the UV posts can be easily reconfigured to allow the approach of an aircraft from another direction from FLS.

If a UV detector with the field of view of 30° is used on an aircraft board, the system sensitivity (range) can be increased up to two times, and the accuracy of determination of the angular directions on the UV beacons can be increased. However, this will narrow the OES navigation zone. The choice of a specific decision should be made taking into account the possibilities of locating equipment in the zone of an airfield. To be able to position at the approach to the airfield, it is necessary to install three beacons, separated from each other at the range of 1–2 km with the cycle of 8–10 km at the trajectory of an aircraft when landing to the point of entry to the glide path [18, 20, 25].

The requirement to prevent collisions with obstacles also means the need to place omnidirectional beacons on pipes, mountain tops and high-rise buildings.

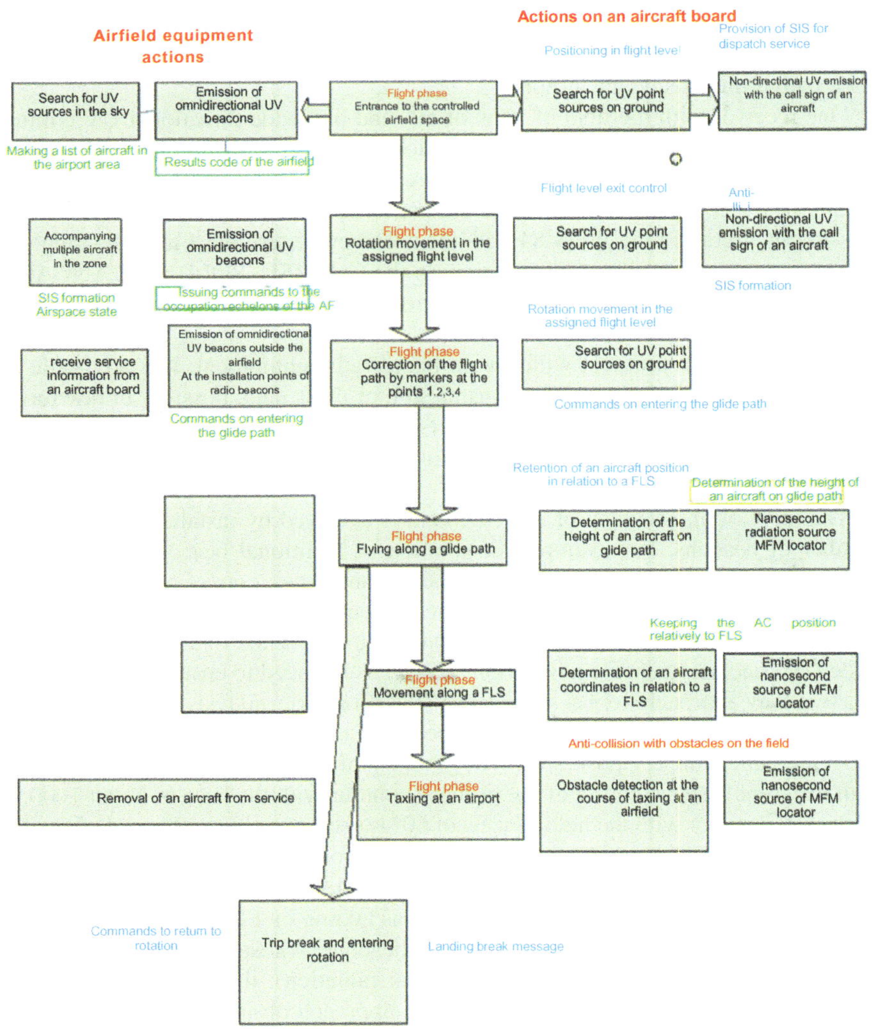

Fig. 1.15 The general scheme of the functioning of the INLS system

The equipment of an aircraft with omnidirectional UV-B beacons makes it possible to detect them by oncoming aircraft. Naturally, in this case, it is possible to detect aircraft from land, as well as to determine their trajectory. The system of 4–5 UV sensors with the field of view of 120° can be installed at an airfield for this purpose. Such system will ensure the detection of an aircraft in the area with the radius of 5–10 km depending on weather conditions and the power of on-board beacons.

For accurate positioning of aircraft and establishing communication with them through a UV channel at extreme ranges, it is possible to use narrow-angle (4°) UV-B

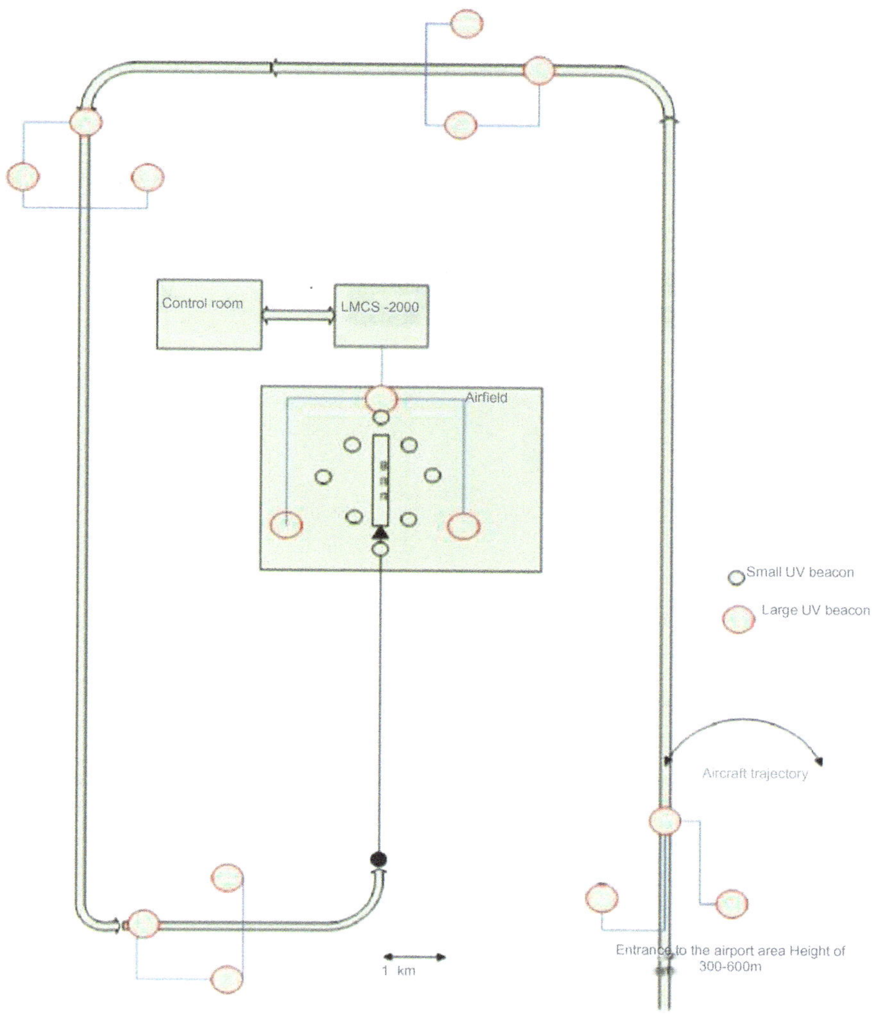

Fig. 1.16 The layout of UV emitters in the vicinity of the airport

sensors. Such sensors will provide the determination of the position of an aircraft with 3 m accuracy at the extreme ranges and the transfer of service information to them along a narrow UV-B beam.

Visibility in fog and clouds provides the potential for informational support for a categorized landing. At the same time, the coordinate accuracy is cardinally improved when approaching a FLS, reaching centimeter values when landing to a FLS. To do this it is necessary to add a few beacons along a FLS route to provide an aircraft with the coordinates when moving along a FLS surface after touchdown.

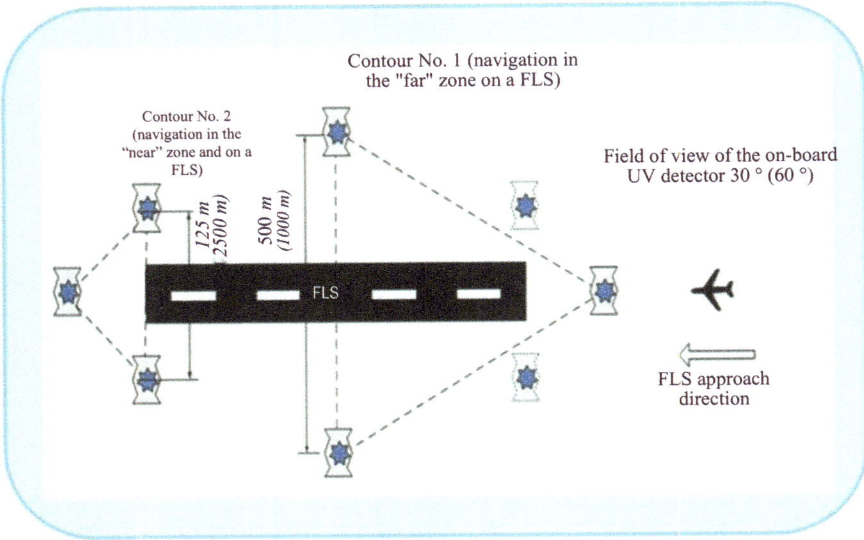

Fig. 1.17 The layout of the UV posts at the territory of the airfield

To control the airspace, it is necessary to place a group of wide-angle receivers at an airfield (at least 4 for the hemisphere observation). In this case, the detection and angle tracking of an aircraft in the vicinity of the airport is carried out in order to form, refine and adjust the aircraft list. It is assumed that the aircraft will be equipped with UV beacons.

Such scheme additionally provides the possibility of determination of not only angles, but also range when an aircraft and an airfield are in synchronous operation.

Besides it is possible to send commands and receive messages between an aircraft and an airport's control room. Thus, it is possible to provide work in case of loss of radio communication (or, if it is desirable to work in radio silence mode).

The expansion of the coverage area of the system is possible by additional placement of receivers and beacons on the route of an aircraft outside the near zone of the airfield [26].

Let us identify the following main variants for construction of advanced vision systems:

1. Enhanced vision systems—EVS. Form an improved image of the external environment from images by thermal imaging (TI) and IR channels and display it on the ILS or MFI.
2. Synthetic vision systems—SVS. Additionally visualize data on the relief, use a database of the ground profile, airports and FLS objects.

3. Automatic vision systems—AVS. Additionally automate the detection of the FLS at landing approach and landing, as well as the detection of obstacles during the taxiing of an aircraft at FLS.

Let's consider the options for building each of them.

The choice of the nearest IR range of 0.7–1 μm allows to apply in this video channel silicon matrices with very high characteristics, for example, matrices with a dynamic range of more than 30,000. Such characteristics of the recording element can potentially allow the formation of an image of the terrain, airfield and FLS with a very high spatial resolution even in low light conditions.

Filtration of visible light (the range less than 0.7 microns) additionally allows to reduce the negative impact of light fog scattered on water particles. However, this channel is only effective in daytime conditions.

In the absence of sunlight, only landing and signal lights can be displayed in the 0.7–1 micron channel image. Under these conditions the thermal imaging channel (8–14 μm) begins to play a significant role, recording the natural thermal radiation of objects and terrain. Further increase in the informativity of the data obtained from these channels can be achieved by joint processing.

As an example, Fig. 1.18 presents the result of such an integrated approach.

Fig. 1.18 The data of the channel of the visible range (left), thermal imaging channel (right) and the result of their joint processing (below)

At night, the images of an airfield infrastructure elements obtained from a thermal imaging channel were additionally contoured by landing lights registered on a visible channel. The available processing tools are sufficient to implement the most complex algorithms for synthesizing various ranges and sequence of frames (equivalent to increase of the sensitivity by two orders of magnitude and the dynamic range by two to three orders).

The use of three-dimensional visualization in the synthesized vision system is as follows. The visualizer is a three-dimensional model of the real world with the ability to display various models with a set position in space. The basis of the virtual world is a model of the earth's surface in the specified area, which with the set accuracy repeats the real terrain, with the applied texture representing the aerial photograph of the area. The virtual world imitates various celestial objects (the Sun, the Moon), displaying their position on the celestial sphere in accordance with the specified date and time [21, 23].

This program allows to display within it the models of various dynamic and static objects (airplanes, ships, ground structures) with reference to geographic coordinates.

This technology allows to trace visually the evolution of dynamic objects in time and space. The areas of use are [27]:

- Visualization and analysis of registration information obtained in field tests by simulating the actual test conditions inside the virtual world;
- In simulation complexes for visualization of the modeling process, including in real time.

The visualizer is built using the Microsoft XNA library, which represents a graphical interface covering. The use of these technologies allows to achieve high performance when drawing a large number of three-dimensional models, which makes it possible to use this module in real-time systems. In actual operation, the performance did not fall below 50 frames per second.

To construct a three-dimensional model of the earth's surface, information on altitudes at the required site is loaded, followed by interpolation and triangulation of the obtained set of points.

The points are stored in an off-line database and can be downloaded from external sources (files, cartographic web servers).

The surface textures are stored as separate files of 256 × 256 pixels and are also stored in an off-line database with the ability to download the missing files from external map servers.

After triangulation the textures are stretched to the resulting grid, resulting in a model of the earth's surface.

The construction of the mesh surface of the terrestrial relief and the surface with an applied texture is shown in Figs. 1.19 and 1.20.

Geographical coordinates are calculated in the WGS-84 system.

The rectangular Cartesian coordinate system is used in the virtual world. The maps use the elliptical or spherical Mercator projection (depending on the map service).

The longitude and latitude are recalculated in the Cartesian coordinate system using the formulas [28]:

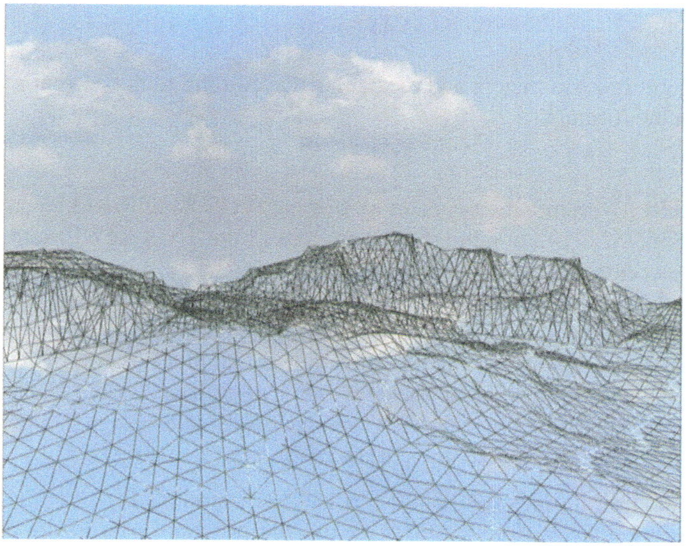

Fig. 1.19 Building of a mesh surface of ground relief

Fig. 1.20 The textured surface

$$X = a \cdot \lambda \tag{1.16}$$

The height is set in meters and recalculated according to the current scale of the maps with the formula:

where:

$a = 6{,}378{,}137, 0$ m—the big semi-axis of the WGS-84 ellipsoid,
$b = 6{,}356{,}752, 3142$ m—the small semi-axis of the WGS-84 ellipsoid,
- is the eccentricity of the earth's ellipsoid,
φ is the latitude of the object in the geographic coordinate system,
λ is the longitude of the object in the geographic coordinate system,
X, *Y* are the Cartesian coordinates at the horizontal plane,
h is the height of the object in the geographic coordinate system,
H is the height of the object in the Cartesian coordinate system,
z is the current map scale (varies in the range from 1 to 20).

For the spherical projection $a = b = 6{,}378{,}137$ is assumed.

The use of off-line storage allows to use the program even in the absence of Internet connection.

For full-fledged autonomous work, it is necessary to first cache the information about the required terrain on the local computer.

The information is stored using the embedded SQLite DBMS, which does not require the deployment of a separate database (DB) server.

Data is stored in one file, which makes it easy to transfer the accumulated information from one computer to another.

The visualizer supports the models in FBX format, which can be obtained by exporting from the 3DSMax program, which allows displaying a wide range of real-world objects.

The coordinates, height and angular position should be specified for an object.

The object will be placed in the real world, taking into account its position on the earth's surface and in relation to other objects (objects which are far from the camera will be covered by nearby objects).

At the same time, inside the visualizer it is possible to build independently a set of graphic primitives with various effects (shading, translucency) to display all kinds of service information [29]:

- Field of view of devices;
- Restriction of different areas;
- Lines, axes, directional signs.

The camera is controlled both in the free mode and with reference to the objects of the virtual world:

- Fixed camera;
- Tracking camera (used to monitor a specific object);
- Directional camera (can be used for imitation of the view from the cockpit or optical device).

For each type of camera, the viewing angle and range of view are adjusted.

The examples of the use of the program to visualize the landing of an airplane at the Sochi airport FLS are shown in Figs. 1.21 and 1.22.

The use of the optical range is always faced with the problem of functioning in adverse weather conditions. In the visible wavelength range cloudiness or fog will disturb terrain visibility.

Fig. 1.21 The use of the program for the visualization of an aircraft landing to the flight landing strip at the Sochi airport

Fig. 1.22 The use of the program for the visualization of an aircraft landing to the flight landing strip at the Sochi airport

In daytime conditions, the scattering of solar radiation in the fog leads to the background, preventing the observation of even bright sources of visible radiation. The proposed use of the UV range has features that distinguish it from the visible and IR ranges. First, even in the daytime, the ozone layer of the atmosphere absorbs the UV radiation of the sun in the range of 200–280 nm, and there are no natural backgrounds at any time of the day. Secondly, in the UV range, there is practically no absorption of radiation by water aerosols and there is only scattering on aerosols, and this is small-angle scattering [18, 21].

As a result, multiple scattered photons retain the direction of movement, and the compactness of the image of a point source is preserved at the passage of sufficient distances in fogs and clouds. This is a non-trivial statement, and in designing it is advisable to rely on direct experimental measurements. Such measurements were carried out repeatedly.

Images of UV-B beacons at FLS can also be displayed in the image, thereby marking the boundaries of the area in which a FLS is located.

Figure 1.23 shows the result of modeling of the visibility of UV beacons by the detector operating in the range of 247.5–287.5 nm, when approaching a FLS during

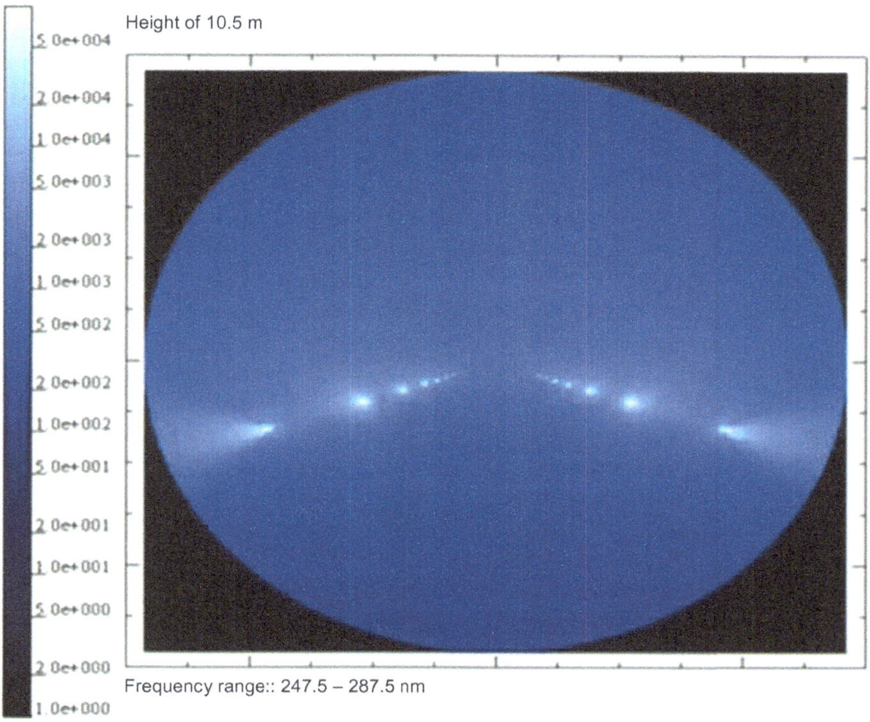

Fig. 1.23 The view from an aircraft from the height of 10.5 m in conditions of very low visibility at contouring of the FLS by UV beacons

the landing process. The simulation was carried out for very low visibility conditions (fog) for an ideal detector.

The result of the simulation clearly shows the efficiency of using UV-B emitters for advanced vision systems to provide visual control of a FLS position during landing.

It is also possible to go to the next stage—the automated vision system, where the task is to show FLS. Here, the ability to transmit data on FLS coordinates through the UV-C channel actually gives a solution to the problem of detecting of FLS and delineating it in the image of the surrounding space, including in conditions of poor visibility.

Thus, the integration of EVS with the UV-C channel data allows, at the first stage, to go to SVS and AVS systems.

To substantiate the currently accepted version of the design of the UV subsystem, we will analyze the type and accuracy of the results of signal registration by photodetectors and their application for aircraft navigation in the near zone of an airport.

Let us suppose that the board is moving with zero roll and tangage angle at a constant speed. In this case, the distance of about 1 km is covered in 5 s. The measurement of angles in relation to an aircraft construction line to the contour No. 1 emitter allows construction of a triangle with one known side and two adjacent angles. This provides calculation of the range to the contour No. 1 with the rather high accuracy of 0.5%. If the measurement of the range by synchronous measurements is done at the same time, the accuracy is increased to units of meters [30].

We note that the presence of data exchange allows you to transfer to an aircraft the coordinates of the emitters of circuit No. 1 and the absolute directions to an aircraft from this point. Thus, the board receives its coordinates in relation to the contour No. 1, and the angular measurements provide information about the roll angle of an aircraft in relation to the plane formed by the axis of an aircraft and the direction to the contour No. 1.

It can be assumed that the use of angular on-board and ground information will allow at observation of a trajectory to calculate the remaining parameters of the board, including the course, roll, tangage, elevation and position in the WGS-84 coordinate system (Figs. 1.24 and 1.25).

The possibilities are greatly extended if two or three UV posts are in the field of view of the board. In this case, the goniometric information to lamps can be used directly to determine the coordinates of an aircraft at known coordinates of the posts.

If the task is "well defined" and the three terrestrial sources together with the board form a stable volume pyramid, angular measurements can provide sufficient accuracy to determine whether an aircraft is at a set flight level.

It is possible to improve significantly the accuracy by observing the trajectory and using data on the rate and straightness of an aircraft flight (or take into account the data on the curvilinearity). The best accuracy can be obtained by interacting with posts, providing the measurement of ranges to posts. In this case, the control of the posts 2 and 3 should be carried out from the post 1, synchronizing the sending of pulses from the lamps of the posts 2 and 3. We note that for the visibility of posts, it is necessary either to set special flight levels so as to ensure the maximum visibility

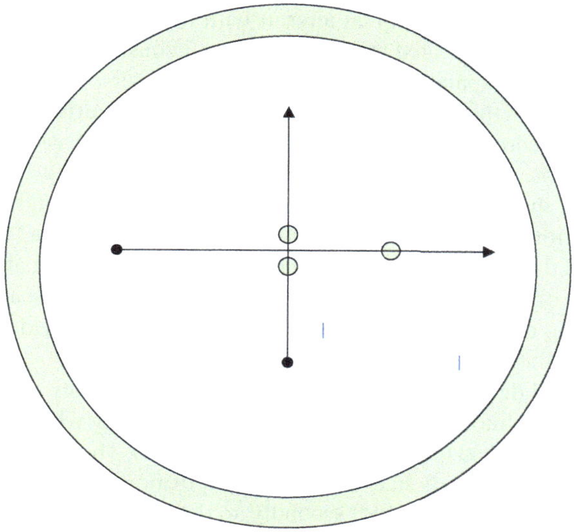

Fig. 1.24 The view of the beacon system from an aircraft at precise glide path moving

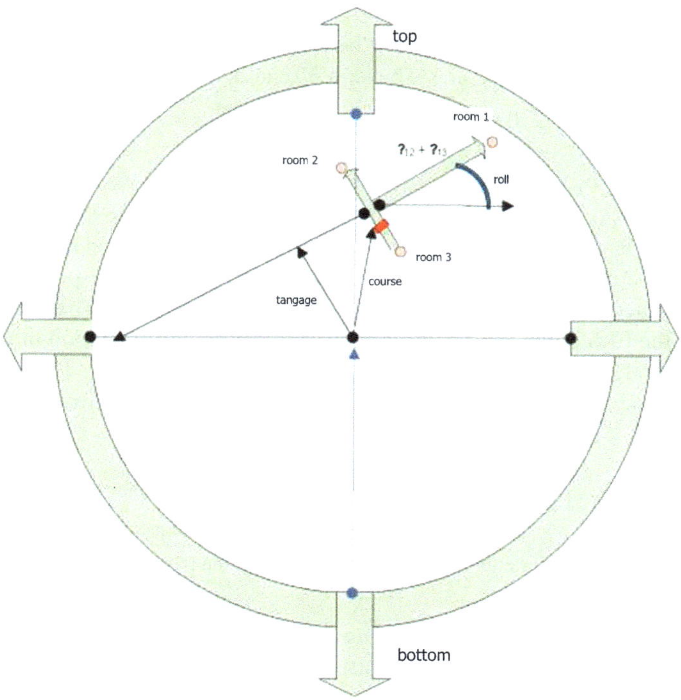

Fig. 1.25 Definition of the course, roll, tangage values

of posts on the route, or to equip an aircraft with more than one photodetector. In fact, two photodetectors located at 45° to an aircraft construction line are enough.

A good decision, expanding the area of controlled airspace, is the location of additional beacons on the ground along the route outside the airfield. The expediency of such obviously correct decisions should be weighed on the basis of economic and other considerations.

We note that the mandatory placement of signal lights on high-rise buildings, pipes and elevations of the terrain can provide such possibility in a "natural" way. It should be noted that such signal lights do not provide the warning about the danger in the presence of fog or clouds during the daytime due to the scattering of solar radiation in a fog or cloud. The proposed UV beacons will provide visibility even with direct solar flash of the photoreceiver with the beacon.

Being on the glide path, an aircraft observes simultaneously the beacons of all three contour posts almost to the distance of decision making when flying according to satellite navigation data (at the level of 1 km from FLS). The presence of the image of the three sources allows, first, to determine the position of the center of gravity of an aircraft in relation to a FLS, and, secondly, to determine the characteristics of an airframe in relation to the plane of an aircraft.

- These measurements make it possible to obtain autonomously on-board the direct measurements that are necessary for a categorized landing. Indeed, the angles vectors between the images θ_{23} = "beacon 2 – beacon 3", θ_{12} = "beacon 1 – beacon2" and θ_{13} = "beacon 1 – beacon 3" can be estimated

  for L = 10 km; $\theta_{23} = 0.014$ rad; $\theta_{12} = 0.04$ rad,
  for L = 5 km; $\theta_{23} = 0.025$ rad; $\theta_{12} = 0.07$ rad,
  for L = 1 km; $\theta_{23} = 0.08$ rad; $\theta_{12} = 0.25$ rad.
  For estimates of θ_{13} value it should be noted that it is close to θ_{12}.

The vector $\boldsymbol{\theta}_{12} + \boldsymbol{\theta}_{13}$ determines the aircraft position in relation to the FLS plane (for qualitative analysis, we assume that the plane formed by the points where the posts are located forms the plane coinciding with the FLS plane) The aircraft roll in relation to the FLS plane can be interpreted as the angle of inclination of this vector in relation to the horizon in the coordinate system of the aircraft. The accuracy of this parameter for a wide-angle receiver is 0.002 rad. This provides the accuracy of roll detection.

For L = 10 km	3°	with the frequency of 5 Hz
For L = 5 km	better than 2°	with the frequency of 100 Hz
For L = 1 km	better than 0.5°	with the frequency over 1 kHz

We note that the statistical processing of instantaneous measurement data over some period of flight significantly increases the accuracy. In any case, this is a rather accurate initial estimate.

The course angle in relation to the FLS beginning can be correlated with the position of the point at the vector $\boldsymbol{\theta}_{23}$, referred by 0.25 from the length of the vector, starting from the beginning of the vector. The accuracy of this direction will be no worse than 0.1°. The tangage angle can be attributed to the displacement of the vector $\boldsymbol{\theta}_{12} + \boldsymbol{\theta}_{13}$ in relation to the construction axis of an aircraft. The accuracy of determination of this angle is also at the level of accuracy of determination of the vector $\boldsymbol{\theta}_{12} + \boldsymbol{\theta}_{13}$.

We note that these definitions are conditional and only emphasize the presence of direct measurements, which can be accurately transformed into control actions taking into account the different accuracy of measurement of different values.

To perform an automatic landing and to ensure the accuracy of positioning and orientation of an aircraft at a FLS, the accuracy at the level of 10 cm is required. This is ensured by positioning of relatively low-energy beacons on the side of a FLS. The location of the beacons at each 200 m is sufficient. Wide-angle photodetectors provide an angular resolution of 2 mrad, which corresponds to a single measurement accuracy of 40 cm in less than 1 m/s. The accuracy of the location measurement to the surface of a FLS is less than 10 cm.

The integral processing of these coordinate-time data received at the rate of more than 1 kHz provides estimation and prolongation of an AC trajectory at the level of centimeter values in relation to FLS.

References

1. The Certificate of flight tests of LMCS-A- 2000 "On the results of the ground-based airfield and flight tests of satellite navigation and landing equipment prototypes of the II and III categories and the system elements on the IL-76 and IL-103 laboratory airplanes under conditions of actual interference of the airfield" No. 163–11-IX. Gromov Flight Research Institute. - 2011.
2. Baburov V.I., Bestugin A.R., Ivantsevich N.V., Kirshina I.A., Sauta O.I., Filin A.D., Shatrakov Yu.G. Comprehensive assessment of navigation information at a priori uncertainty. - Issues of radio electronics. 2016, No. 6, p. 66–70.
3. GOST 26121–84. Airborne instrumental approach systems for landing. Terms and definitions. — M.: Standards Publishing, 1984. — 8 pages.
4. N.I. Didenko, B.P. Eliseev, O.I. Sauta, A.Yu. Shatrakov, A.V. Yushkov. Radio technical support of military and civil aviation flights is a strategic problem of the Arctic zone of Russia. – The scientific proceedings of MSTUCA, V. 20, No. 5, 2017, p. 8–19.
5. Zavalishin O.I. Local functional extensions to GNSS "RDPF Spektr" // Navigation news. – 2009. - No. 4 – p. 25–31.
6. Zavalishin O.I. Cleared to land // Aerospace Courier. - 2011. - No. 2 – p. 71.
7. Zatuchny DA (2010) Assessment of the degree of influence of various factors on the navigation definitions of aircraft with the use of SRNSS. Scientific Bulletin of the Moscow State Technical University of Civil Aviation. 159:143–147
8. Zatuchny D.A. To the problem of sustainability and radio safety of navigation systems of civil aircraft using precise satellite navigation systems. Proceedings of the international symposium Reliability and quality. 2018. V.1. P. 38–40.
9. Karutin S.N. Differential correction and monitoring of Global Navigation Satellite Systems (GNSS) / S.N. Karutin, I.B. Vlasov, V.V. Dworkin // - M .: Moscow State University Press, 2014. – 464 pages.

10. Proceedings of the 12th ICAO Air Navigation Conference (2012, November 19–30) / Montreal, Canada.
11. On the use of the global navigation satellite system GLONASS in the interests of the socio-economic development of the Russian Federation: The decree of the President of the Russian Federation from 17.05.2007 No. 638 // Russian newspaper. - 2007, 23 May. - No. 4370[70].
12. About Global Navigation Satellite System GLONASS: The order of the President of the Russian Federation from 18.02.1999. № 38-rp // Garant. Information and legal support [Electronic resource] - https://base.garant.ru/180169/.
13. Patent on an invention. 2441203 RF. Integrated universal all-weather method for determination of the location and landing of an aircraft and device for its implementation / Zavalishin O.I. – Declared on 29.09.2010; Published on 27.01.2012, Bul. No. 3
14. Patent on an invention. 2542325 RF. The method of location of an aircraft / Zavalishin OI – Declared on 24.07.2013; Published on 20.02.2015, Bul. No. 5
15. Patent on an invention. 2393504 RF. The method of determination of the unacceptable anomalies of the received signals of navigation satellites and the device for its implementation / Zavalishin O.I. – Declared on 23.10.2008; Published on 27.06.2010, Bul. No. 18
16. Patent on an invention. 2541691 RF. The way to improve the integrity of the output signals of on-board satellite navigation receivers / Zavalishin O.I., Lebedev B.V. – Declared on 12.08.2013; Published on 20.02.2015, Bul. No. 5
17. Patent on an invention. 2653066 RF. The method of determination of the ionospheric storm using ground stationary LMCS / Zavalishin O.I. – Declared on 29.08.2017; Published on 07.05.2018, Bul. No. 13
18. Povalyaev A.A. Satellite radio navigation systems: time, clock readings, formation of measurements and determination of relative coordinates. – M.: Radio engineering, 2008. – 328 pages.
19. Volume 1. Radio navigation equipment. Annex 10 to the Convention on International Civil Aviation: Aviation telecommunication. International Civil Aviation Organization, 2006. – 616 pages.
20. Rubtsov V.D. The method of calculation of distributions of interference and its mixture with a signal using experimental curves of the distribution of the envelope of the interference // Informatization and communication. – 2015. - No. 2. – p. 57–61.
21. Soloviev Yu.A. Satellite navigation systems. – M.: Eco-Trendz, 2000. – 270 pages.
22. Soloviev Yu.A. Satellite navigation and its applications. – M.: Eco-Trendz, 2003. – 326 pages.
23. Federal Aviation Rules "Air Traffic Management in the Russian Federation". Approved by the Order of the Ministry of Foreign Affairs of the Russian Federation from 25.11.2011 No. 293 / Garant. Information and legal support [Electronic resource] - https://base.garant.ru/70117238/.
24. Federal Aviation Rules "The procedure for the implementation of radio communications in the airspace of the Russian Federation." Approved by the Order of the Ministry of Foreign Affairs of the Russian Federation from September 26, 2012 No. 362 / Consultant Plus [Electronic resource] - https://www.consultant.ru/document/cons_doc_LAW_145093/.
25. Federal Aviation Regulations "Radio technical support of aircraft operations and aviation telecommunications in civil aviation" approved by the Order of the Ministry of Foreign Affairs of the Russian Federation from 20.10.2014 No. 297 / Garant. Information and legal support [Electronic resource] - https://base.garant.ru/70812462/.
26. ICAO: International Standards and Recommended Practices. ANNEX 10. vol. 1 (Radio Navigation Aids), 1999–2006.
27. Sharov V.D. Flight safety risk management methodology at the airline level: Thesis for the degree of Doctor of Technical Sciences. Speciality 05.22.14 - Operation of air transport. – M.: MSTUCA, 2016 – 398 pages.
28. Maslov V.Yu. Anomalous errors of satellite navigation systems in determination of the location of an aircraft in mountainous terrain and their elimination with the use of radio polarimetry methods: Thesis for the degree of Doctor of Technical Sciences. Speciality 05.22.13 - Navigation and air traffic control // V.Yu. Maslov – M.: MSTUCA, 2007 – 260 pages.

29. Nechaev E.E. Improving the accuracy and reliability of measurements in integrated satellite navigation systems using the double averaging method: / Thesis for the degree of Doctor of Technical Sciences. Speciality 05.22.13 - Navigation and air traffic control // E.E. Nechaev – M.: MSTUCA, 2011 – 183 pages.
30. Sobolev S.P. Methods for assessing and monitoring of the integrity of the on-board navigation equipment of the satellite landing system: thesis for the degree of candidate of Technical Sciences: 05.12.14. - St. Petersburg, 2007. - 179 pages.

Chapter 2
Assessment and Reliability Assurance of the Navigation Systems of the Civil Aircraft Under the Conditions of Influence by Different Interference

2.1 Method of Calculations Confirming the Performance and Reliability of the GBAS Station

For the reliability assurance of the GBAS satellite landing system (SLS) the software and hardware implementation in the form of a dual set with "hot" redundancy is proposed. Moreover, both sets of SLS (GBAS) equipment constantly (at least 1 Hz) "interrogate" each other about the working order of the second set.

The SLS (GBAS) has a built-in automatic control system to monitor the status of individual station devices, provide disconnection of faulty devices and automatically connect redundant ones, as well as to generate and issue data on the current state of the station's operability. This solution significantly improves the reliability, continuity and availability of the navigation and landing system [1].

When a half-set fails or even when it is lost, the station deteriorates its characteristics, but it remains operational and provides the implementation of the precision approach function. To implement the technology of "hot" redundancy switch variant has been developed in the SLS (GBAS) equipment to provide the switching of the power amplifier of each of the half sets to one VHF antenna. Antenna-feeder devices are made of thick-walled material that allows to withstand (using spreaders) the wind load of the air flow at wind speeds up to 50 m/s, as well as operation in conditions of unheated rooms, cars, etc.

To ensure the reliability and stability of the SLS (GBAS) station masts, there is the option of closing with three-side guy-wires separated at an angle of 120° in relation to each other and 300 in relation to the vertical telescopic mast.

To assess the SLS (GBAS) reliability indicators during the trial operation the confidence intervals method was used. The main parameters of the landing systems influencing safety are integrity, continuity of service and availability. All these parameters are directly dependent on the reliability of the system [2, 3].

O. I. Zavalishin et al., *Modern Requirements for Noise Immunity Aircraft Navigation Equipment*, Springer Aerospace Technology,
https://doi.org/10.1007/978-981-16-0073-9_2

In the process of developing and certifying the system, the equipment underwent several types of tests (factory, acceptance, certification and operational) in which all system parameters should be confirmed.

The Annex 10 of the ICAO SARPs establishes the need to confirm the theoretically justified reliability parameters during the operation of the system.

Thus, the confirmation of the reliability of the system is an obligatory task of the GBAS scientific and technical support (LMCS).

For one non-redundant SLS set implementing landing on I, II and III categories of ICAO, the mean time between failures (MTBF) is set not less than 2000 h at the 90% confidence interval. In addition, reliability tests should last at least a year to take into account the effect on the system of seasonal changes, that is, weather conditions, as well as power sources, preventive maintenance, and so on.

The cause of failures (cessation of signal emission) of SLS (GBAS) during operation may also be external factors, such as interference with the emission of radio engineering systems, multiple reflections or significant precipitation in the form of snow. These failures should be clearly identified and excluded when assessing the reliability of the system.

The suitable method for assessing the reliability of a particular tool is registration and calculation of the average mean time between work breaks (MTBW) value based on the latest 5–8 equipment failures. For MTBW values defined in ICAO Annex 10 (2000–4000 h), the procedure for confirmation of a theoretically reasonable value may take several years, which is highly undesirable from the point of view of system approval for operation, especially a fundamentally new system, which is SLS (GBAS).

ICAO allows the accelerated MTBW confirmation by testing several identical systems operating under the same conditions. In our case, for assessing GBAS (LMCS) MTBW it is advisable to use the methodology developed by the AWOG certification subgroup (the group of experts of the European Bureau of ICAO for all-weather flights).

The accelerated service continuity assessment methodology is based on the assumption that if the developer can provide the rationale that the theoretical MTBW value is two times higher than the specified value, then the service continuity assessment can be carried out at the 60% confidence interval. If the condition of double excess is not met, the confidence interval should be 90%. For the case of SLS (GBAS), the calculated MTBW value is equal to 6042 h, which is three times the specified value [3, 5].

The methodology involves the determination of the boundaries within which the characteristics of the equipment are either acceptable or not. These boundaries are based on the following parameters [6, 7]:

- User's risk (β): The probability of accepting equipment with the MTBW value equal to the lower limit (θ_1 MTBW value that is not applicable for this type of equipment), that is, the probability of accepting equipment with the MTBW less than θ_1 is equal to β;

- Manufacturer's risk (α): The probability of rejection of equipment with MTBW equal to the upper limit (θ_0), that is, the probability of accepting equipment with the MTBW less than θ_0 is equal to α;
- Discriminatory ratio (d) is the ratio of the upper limit to the lowest $d = \theta_0/\theta_1$.

When testing, the value of θ_1 is assumed to be equal to the MTBW value given for this equipment.

The technique sets the following values of the above parameters (Table 2.1).

These parameters define the duration of the test, which allows to make the decision on the compliance of the set MTBW value, measured depending on the number of registered failures (Table 2.2).

These boundaries for MTBW of 2,000 and 4,000 h are shown in Figs. 2.1 and 2.2. This takes into account that to assess the influence of local airport conditions, the minimum test period according to the Amendment 74 of the Annex 10 of ICAO should be one year (8,800 h).

Although the above programs are designed to demonstrate that the MTBW corresponds to the set value of 60% ($\beta = 0.4$) confidence interval, in reality, the confidence interval is higher and its value depends on the point at which the decision to accept the system is made (the intersection point of the failure curve with the lower limit of decision-making area).

So, if the equipment is accepted at a small number of failures, the confidence interval corresponding to the specified MTBW value will be much higher.

For example, if equipment is accepted at zero failure quantity (after a year of testing), the confidence interval will be 98% and 88% for MTBW of 4,000 and 2,000 h, respectively (Figs. 2.3 and 2.4).

After initial certification, it is necessary to continue monitoring of the reliability of the installed systems. The suitable method for assessing the reliability of the specific

Table 2.1 Parameter values

User's risk (β)	0.1
Manufacturer's risk (α)	0.4
Discriminatory ratio (d)	2

Table 2.2 The number of registered failures

Number of failures	Acceptance time of θ_1	Rejection time of θ_1
0	1.62	N/A
1	3.01	N/A
2	4.39	N/A
3	5.78	1.15
4	6.30	2.54
5	6.30	3.92
6	N/A	6.30

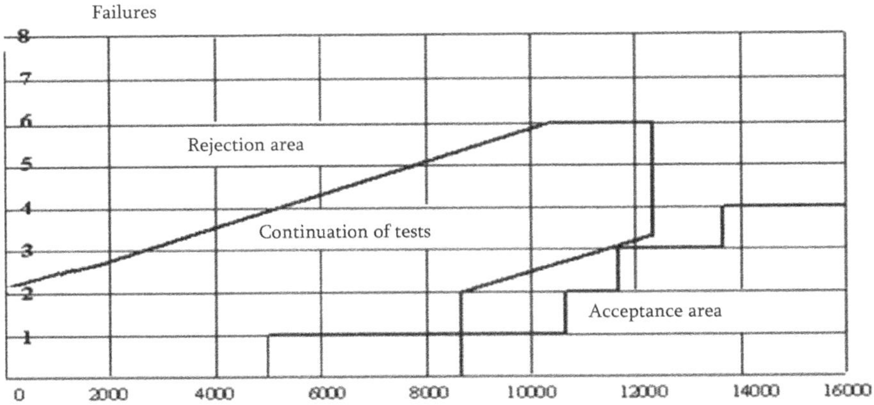

Fig. 2.1 The decision limits for the 60% confidence interval at the MTBW of 2,000 h

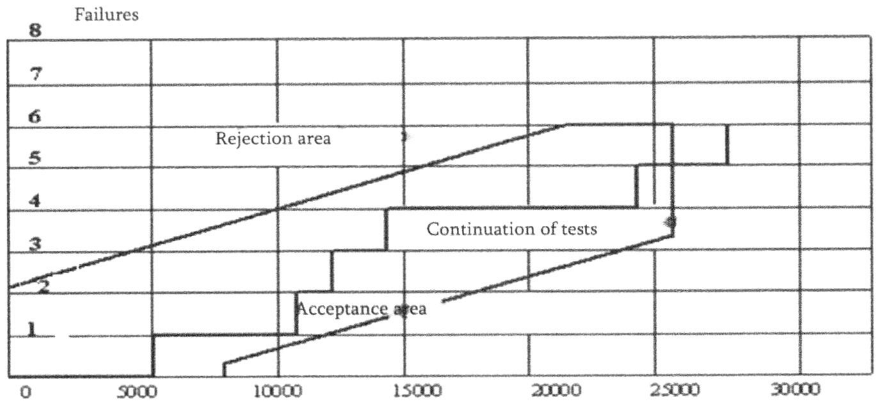

Fig. 2.2 The decision limits for the 60% confidence interval at the MTBW of 4,000 h

system's MTBW is registration and calculation of the average MTBW (mean time between work breaks) value based on the latest 5–8 failures.

The decision-making rule may provide that if the point estimate falls below the required MTBW value, then the equipment is updated. During the certification period it is recommended to carry out a regular, at least once a month, analysis of MTBW. The analysis should reveal whether failures are systematic or random. The reason for systematic failures must be identified and excluded [8].

When testing a group of systems for the case of several almost identical systems operating under the same conditions, the assessment can be carried out on the basis of the total amount of work of all systems. In this case, the time required to demonstrate the reliability of the system can be significantly reduced. The main condition for this is the identity of all systems and their operating conditions.

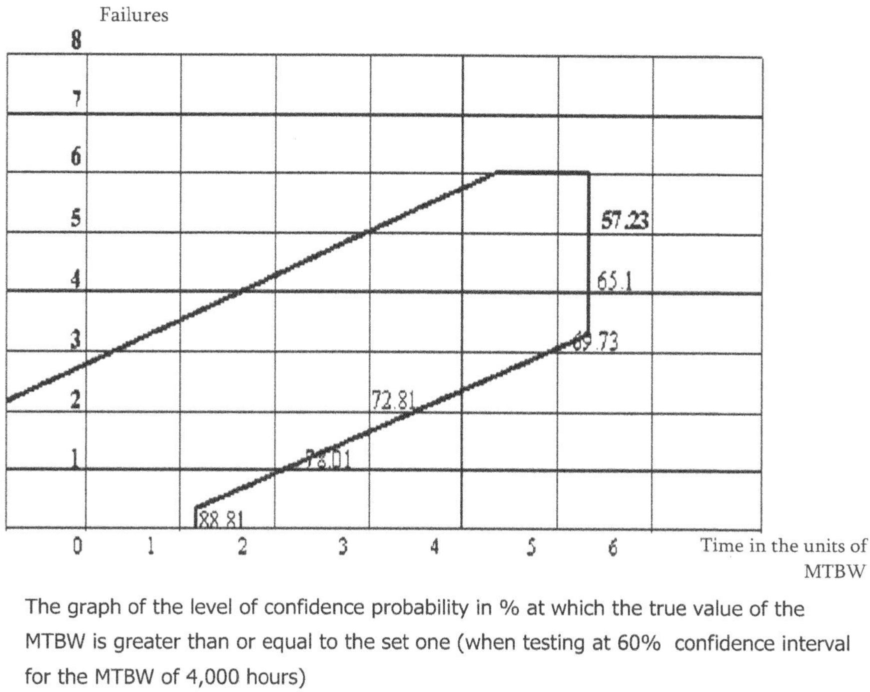

Fig. 2.3 The confidence level graph (when tested for the MTBW of 4,000 h)

An example of a test program of five SLS (GBAS) landing systems at the four levels (MTBW of 4,000 h) that depends on the real time is shown in Figs. 2.5 and 2.6.

For the GBAS course beacon of the III category with MTBW of 8,000 h the test duration is 42,000 h with 90% confidence level. In case of installation of systems according to Fig. 2.7, the equivalent time is accumulated after approximately 12,000 h. In group testing of systems, there is a risk that the procedure of reliability demonstration will be "contaminated" by an uncharacteristic system that will create emissions in statistics.

It is important to exclude this system from the group of systems being tested. But in equal measure the system should not be excluded from the group, if the non-standard performance characteristics of the system simply represent a statistically rare event.

It is assumed that a large variation in the MTBW observation on the same type of systems is due to a small number of failures, so it is desirable to have a mechanism for excluding the system from the group if the deviation of this system from the others is much higher than expected. Such a mechanism consists in applying a test for discrepancy of characteristics [9, 10].

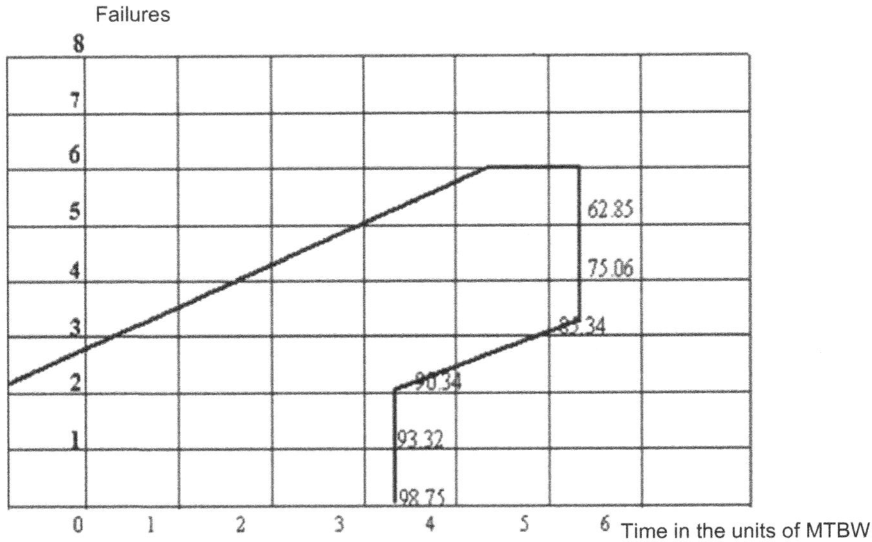

The graph of the level of confidence probability in 60% which the true value of the MTBW is greater than or equal to the set one (when testing at 60% confidence interval for the MTBW of 2000 hours)

Fig. 2.4 The confidence level graph (when tested for the MTBW of 2,000 h)

This test demonstrates whether systems have significantly worse reliability characteristics than others, which raises the question of assuming that the probability distribution is homogeneous between systems within the same group.

The sequential (logical) test for discrepancy can be used for continuous analysis to an extent the characteristics of one system differ from the characteristics of the other systems in the group such that its exclusion is justified.

The exact decision regarding the discrepancy in the characteristics of the systems can be obtained on the basis of modeling of the real situation. In the case of detection and elimination of systems with non-standard characteristics, this type of system should be considered as inclined to deviations, which may cause any consequences for the time required for the introduction of subsequent systems of this type.

If at the certification stage there is a border crossing into the "rejection" area and if there is an assumption that this is caused by a system with uncharacteristic characteristics, then the tests can be continued. In this case the decision regarding the acceptance of a system can be made only when the test for discrepancy reveals a system with uncharacteristic and when the resulting failure curve at the exclusion of a non-standard system first crosses the boundary of the "acceptance" area [11].

The introduction of a subsequent system does not require re-certification, as for a new system, it is almost identical to the type of previously certified systems, and the time of its introduction is significantly reduced. The requirement of reliability

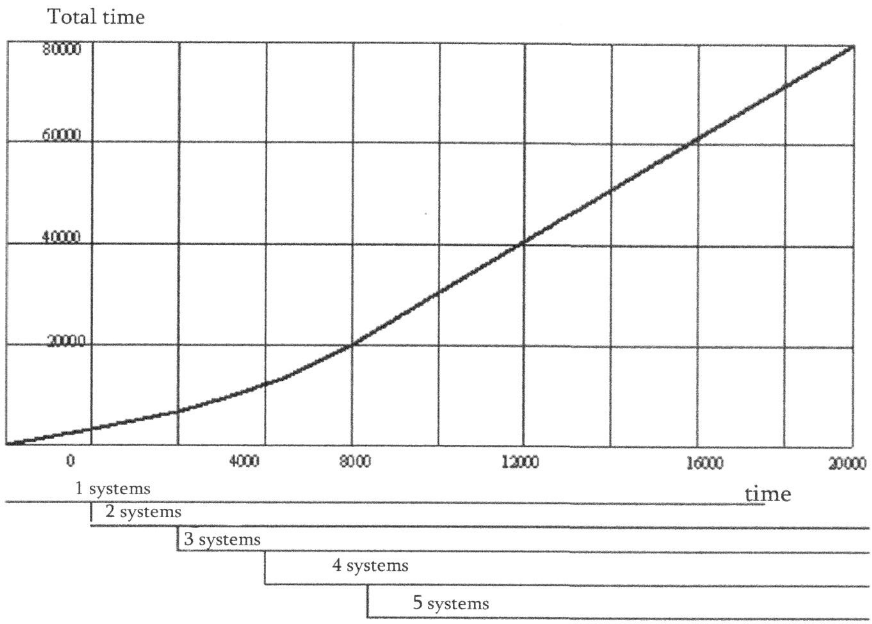

Fig. 2.5 The total time for tests of five systems

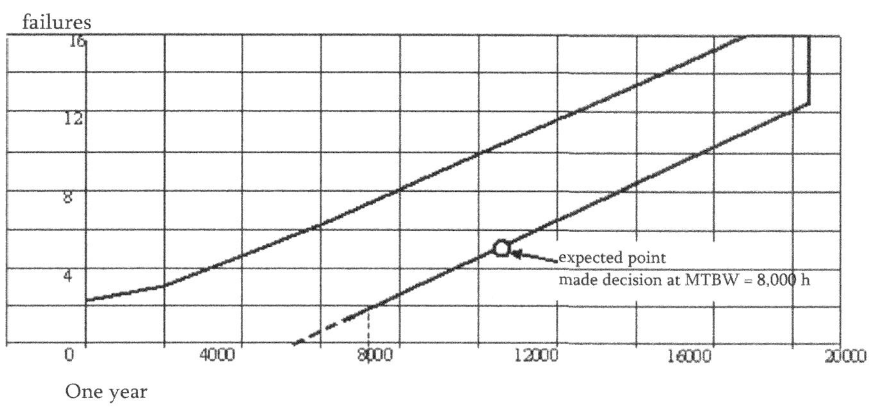

Fig. 2.6 Tests of five SLS landing systems at the four levels

demonstration becomes in this case the requirement of demonstration of its correct installation. Such a procedure can be carried out for about a month.

The accelerated certification program of the subsequent system is determined by the assessment of whether the system is non-standard. This decision is made on the

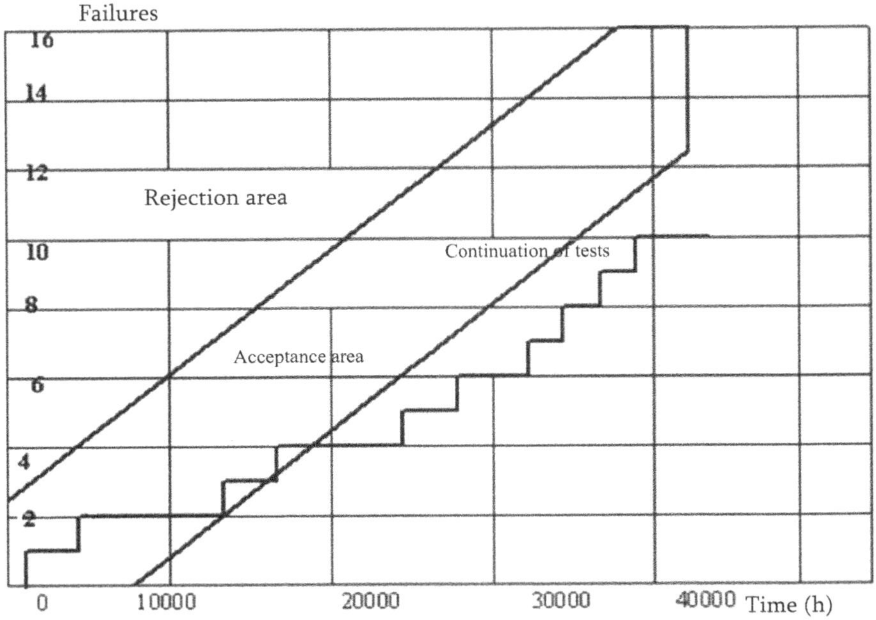

Fig. 2.7 The decision limits at 90% confidence interval in the MTBW of 2,000 h

basis of a consistent discrepancy test carried out during the certification process of a group of systems.

If the type of systems is not defined as inclined to the demonstration of non-standard characteristics, the implementation of subsequent systems can be carried out under an accelerated program.

The subsequent system certified under the accelerated program in the initial period should be subjected to more severe post-certification monitoring.

If it is impossible to carry out tests using the accelerated method (at 60% confidence interval), then at 90% confidence interval, the test time increases significantly (Table 2.3 and Fig. 2.7).

From the example of this graph it can be seen that the acceptance of the system can be carried out after 17,120 h with three failures. Using the Table 2.4, we get the lower limit of MTBW at 90% confidence interval equal to $1.178 \times 2{,}000 = 2{,}356.6$ h, which is the requirement for MTBW of 2,000 h when 90% confidence interval is met.

In Figs. 2.8 and 2.9, there is a significant increase in the expected testing time compared to testing at 60% confidence interval. So for the true MTBW value equal to the required one, the expected time is $7.8\theta_1$ or approximately 31,000 h for the equipment with MTBW = 4,000 h.

This methodology with the examples of MTBW estimates and the estimated MTBW value (6,042 h) for a non-redundant system allows to conclude that the

Table 2.3 Test duration limits

Number of failures	Acceptance time of θ_1[a]	Rejection time of θ_1
0	4.40	N/A
1	5.79	N/A
2	7.18	N/A
3	8.56	0.70
4	9.94	2.08
5	11.34	3.48
6	12.72	4.86
7	14.10	6.24
8	15.49	7.63
9	16.88	9.02
10	18.26	10.40
11	19.65	11.79
12	20.60	13.18
13	20.60	14.56
14	20.60	15.94
15	20.60	17.34
16	N/A	20.60

[a]In the units of MTBW

Table 2.4 90% lower confidence limit after acceptance

Number of failures	Testing time[a]	90% lower confidence limit
0	4.40	1.9109
1	5.79	1.4311
2	7.18	1.2633
3	8.56	1.178
4	9.94	1.1278
5	11.34	1.0956
6	12.72	1.0734
7	14.10	1.0575
8	15.49	1.0459
9	16.88	1.0372
10	18.26	1.0305
11	19.65	1.0254
12	20.60	1.0194
13	20.60	1.0045
14	20.60	0.9835
15	20.60	0.9613

[a]In the unzts of MTBW

Fig. 2.8 The dependence of the acceptance of the system from the real MTBW

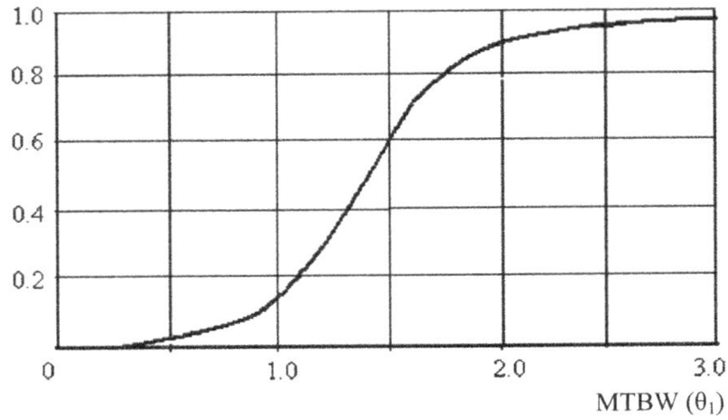

Fig. 2.9 The dependence of the expected testing time on the real MTBW

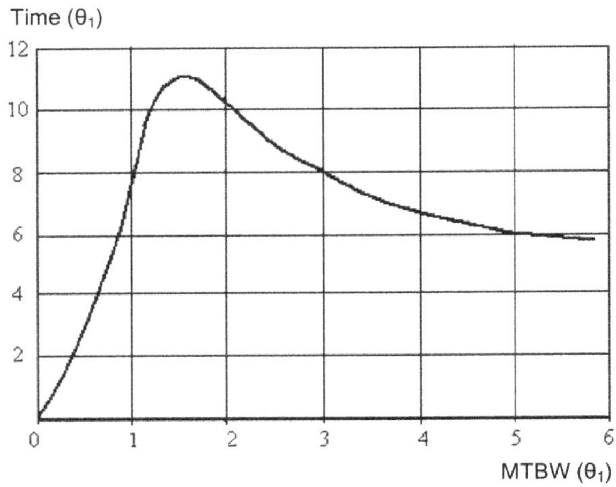

MTBW for redundant SLS (GBAS) will be significantly higher than the specified value (10,000 h). The mean time between failures is 40,332 h, which is significantly higher than the estimated value [10, 12].

Reliability can be achieved through the use and regulation of many factors. For specific conditions (in a certain condition of the environment, an optimal selection of factors is necessary), low reliability can be increased by redundant elements of the system, that is increase in staff. The more rational way is the inclusion in the system of components with higher parameters of MTBF.

The following formula expresses the reliability of the equipment in percent [13, 14]:

$$\mathrm{R} = 100\,\mathrm{e}^{-t/m}, \tag{2.1}$$

where R is reliability (probability of equipment operability in specified parameters during the set time t); e is the base of the natural logarithm; t is the set period of time corresponding to this operation; m is the equipment mean time between failures for a non-reserved set of elements.

Obviously, reliability rises with the mean time between failures, and MTBF is another more convenient way of expressing reliability.

Also, the increase in the mean time between failures implements the required level of reliability. The MTBF characteristics affecting equipment reliability include [15]:

(a) The structural reliability of the station;
(b) The depth of redundancy;
(c) The reliability of auxiliary devices;
(d) The proper operation;
(e) The external influencing factors, such as temperature, humidity, voltage surges, etc.

The SLS mean time between failures in operation $MTBF_{LMCS}$ according to the data from Table 2.4:

$$MTBF_{ЛККС} = \frac{\Sigma T \text{ наработки комплектов ССП (GBAS)}}{\text{Число отказов комплектов ССП (GBAS)}} = \frac{80664}{2} = 40332 \text{ часов} \quad (2.2)$$

The equipment availability factor (A) as a percentage can be defined as the quotient from the actual work time to the assigned work time, taken over a long period, i.e., as follows:

$$A = \frac{\textit{Фактическое время работы комплекта ССП (GBAS)}}{\textit{Установленное время работы комплектов ССП (GBAS)}} \times 100\%. \quad (2.3)$$

The set work time is the sum of the actual work time and time of work breaks for scheduled maintenance and/or replacement of failed equipment.

To ensure high level of equipment availability, the following factors are important [16]:

(a) The equipment reliability;
(b) The quick response of maintenance personnel to equipment malfunctions;
(c) The technical staff qualifications;
(d) The replacability of replaceable modules;
(e) The availability of replaceable modules and components;
(f) The provision of relevant test equipment;
(g) Spare equipment and/or facilities.

We present the calculation of reliability indicators for typical operations.

For oceanic and remote regions, inland routes, airfield area and non-precision landing approach [17]:

$R = 100\ e^{-t/m} = 100\ e^{-1/40332} = 100\ e^{-0.0000248} = 99.9975\%$

$t = 1$ h.

m = 40,332 h.

For precision landing approach of the CATI:

$R = 100\ e^{-t/m} = 100\ e^{-0.0083/40332} = 100\ e^{-0.000000206} = 99.9999999\ldots\%$
t = 30 s = 0.0083 h.
m = 40,332 h.

For precision landing approach of the CATII and CATIII:

$R = 100\ e^{-t/m} = 100\ e^{-0.00415/40332} = 100\ e^{-0.000000103} = 99.999999\ldots\%$
t = 15 s = 0.00415 h.
m = 40,332 h.

We give the calculation of the availability factor for a non-redundant set.

$$A = \frac{\text{Фактическое время работы комплекта ССП (GBAS)}}{\text{Установленное время работы комплектов ССП (GBAS)}} \times 100\% = \frac{40332}{40334} \times 100 =$$
$= 99{,}995\%.$

The actual working time of the SLS (GBAS) sets = 40,332 h.

Set working time of the SLS (GBAS) sets = Actual working time of the SLS (GBAS) sets + Break time for scheduled maintenance and replacement of failed blocks = 40,332 + 2 = 40,334 h.

Break time for scheduled maintenance = 0, since maintenance was performed without shutting down the stations.

Time to replace failed blocks = 1 h (Ostafyevo) + 1 h (Samara) = 2 h.

We present the calculations confirming the efficiency of the OES [15, 17].

The requirements for positioning accuracy at landing can be implemented by varying the geometry of the location of the beacons. At the same time, it is necessary to proceed from the assessment of the accuracy of determination of the coordinate (range):

$$\Delta L = \frac{L^2}{D}\Delta\theta, \tag{2.4}$$

where the angle error is given in radians.

Here L is the distance from an aircraft to a beacon; D is the base distance between the beacons; Θ is the angle between the images of the beacons on the sensor.

We note that the accuracy of measurement of angles between the images on the UV sensor of two beacons depends on several factors:

- Field of view of UV sensor: Θ_{sensor}
- Sensor quality: K_{sensor}
- Statistics $\sqrt{N}$, where N is the number of photons by which the value of L is estimated:

$$\Delta\theta = \frac{K_{\text{сеенсор}}}{\theta_{\text{сенсора}}}\sqrt{N}, \tag{2.5}$$

We note that in the production of sensors there are conflicting requirements between measurement accuracy, sensitivity, weight, size and cost. These relations are determined both by the general laws of physics and by the current technological level.

Currently, the following estimates for the sensors are true.

For $\Theta_{\text{sensor}} = 60°$

$$\Delta\Theta = 0.5/\sqrt{N}, \text{degrees}$$

For $\Theta_{\text{sensor}} = 30°$

$$\Delta\Theta = 0.2/\sqrt{N}, \text{degrees}$$

For $\Theta_{\text{sensor}} = 4°$

$$\Delta\Theta = 0.05/\sqrt{N}, \text{degrees}$$

From these estimates it is possible to make calculations of configuration parameters that meet the requirements. In this case, it is possible to estimate the requirements for the intensity of UV emitters. When carrying out the assessments, it should be taken into account that micronavigation systems are built into the UV sensors, allowing them to separate themselves from the evolution of the aircraft at times of about 1 s. Therefore, the estimates for N can be considered as the intensity of the recorded signal from the UV beacon [18].

The ideology of building of the system is based on the transmission of beacon coordinates and service information through the UV channel. The transmission rate depends on the intensity of the received signal. The data rate estimates depend on the modulation method and transmission quality requirements.

The following experimental data can be taken as the standard estimate: transmission with the rate of 1500 bytes per second is carried out at the signal intensity of more than 15,000 photons per second.

At extreme ranges wide-angle systems provide transmission at 100 bps, and narrow-angle systems are the next larger.

We also note the implemented functions:

- Signal gating when the intensity exceeds that transmission in real time by computational means. This extends the dynamic range by several orders of magnitude and allows to record both beacons at extreme ranges, and when moving along the FLS;
- Consecutive turning on of beacons due to a single time provided by STM signals. This property allows to record the coordinates of signal points in the presence of

their spreading in fog. In addition, this function provides the correct transmission of messages from different beacons in real time (time division).

Let us consider the typical values of the ranges characteristic for the execution of the gating tasks.

L = 1000 m, D = 500 m (typical for an airfield) is the distance to make the decision about entering the second circle.

For $\Theta_{sensor} = 60°$

$$\Delta L = 17/\sqrt{N}, \text{ meters.}$$

30 photons are sufficient to achieve the requirement of 3 m. We note that the typical value of a signal from a typical emitter will be $N = 10^4$ photons per second at the range of 1 km. This ensures the set accuracy with the frequency of 300 Hz. An accuracy of 1 m will be provided on 300 photons with the frequency of about 30 Hz. The requirements are met with a wide margin, and for the range of 1 km, it is possible to use any of the listed angular values of the sensor.

The range L = 2,000 м.

For $\Theta_{sensor} = 60°$

$$\Delta L = 72/\sqrt{N}, \text{ meters.}$$

600 photons are necessary to achieve the requirement of 3 m. The typical value of a signal from a typical emitter will be $N = 10^3$ photons per second at the range of 2 km. This ensures the set accuracy with the frequency of about 2 Hz.

The note on the possibility of use of micronavigation provides meeting the requirement:

The range L = 3,000м.

For $\Theta_{sensor} = 60°$

$$\Delta L = 150/\sqrt{N}, \text{ m.}$$

2500 photons are necessary to achieve the requirement of 3 m. A typical value of a signal from a typical emitter will be N = 300 photons per second at the range of 3 km. This ensures the set accuracy with the frequency of about 0.1 Hz.

The note on the possibility of use of micronavigation provides meeting the requirement. However, this is already within the limits of the system's capability and there is no margin for adverse weather conditions. Therefore, in this case, it is advisable to increase the power of the emitter (by 10 times) or go to the angle of view of 30°.

For $\Theta_{sensor} = 30°$.

$$\Delta L = 60/\sqrt{N} \text{ meters.}$$

400 photons are sufficient to achieve the requirement of 3 m. We note that the typical value of a signal from a typical emitter will be N = 300 photons per second at

the range of 1 km. This ensures the set accuracy with the frequency of 1 Hz (taking into account the use of micronavigation).

The use of an emitter with the intensity increased by 10 times provides an energy reserve in case of adverse weather conditions.

Thus, for the positioning range of up to 3 km from a FLS with the accuracy better than 3 m, it is advisable to use UV sensors with the field of view of 30°. UV emitters should be used with increased power.

The calculated reliability indicator is the mean time between failures. At the same time, the intensity of the flow of failures was preliminarily separately calculated for the components of the complex—the blocks and the component of the blocks—their modules [19].

The reliability calculation was made by the method of forecasting according to GOST 27.301–95 using data on the reliability of components, presented in reference publication 22 of the Central Scientific Research Institute "Reliability of radio and radio products", RD V 319.01.20–98. The system consists of a receiver of infrared radiation, a video camera of the optical range, built on a matrix with the resolution of 5.5 Mpix, and an optoelectronic system operating in the UV range.

The mean time between failures of video cameras built on matrices with a resolution of 5.5 Mpix is 50,000 h (20×10^{-6} h^{-1}).

The mean time between failures of modern photodetectors for thermal imaging equipment is close to 25,000 h (the flow of failure is 40×10^{-6} h).

2.2 The Method of Reducing the Influence of Tracking Failures of the Aircraft Positioning, Arising as a Result of Fast Maneuvering When Flying in Mountainous Areas and at Low Altitudes by Optimizing the Tracking System on Speed-of-Response

When maneuvering aircraft the shading of NSC with elements of an aircraft structure is possible. This may be one of the reasons for the failure of tracking of an aircraft flight. In addition, the cause of the failure of tracking can be a flight in an urban or mountainous area. One of the possible prevention of tracking failure is a complex processing at the expense of the remaining elements of the working constellation. However, in some cases, for example, when the antennas are completely shadowed by the airframe or the navigation equipment fails, the failure of tracking of the SRNS signals still occurs. For this problem, it is necessary to determine quickly the aircraft location under such circumstances.

For the determination of RMSD of an aircraft acceleration, it is necessary to use the following formula:

$$\delta_A^2 \approx \alpha^2 v_0^2 \left(\frac{\delta_\varphi}{57,3} \right)^2, \tag{2.6}$$

where 57.3 is the value connected with the conversion from degrees to radians, δ_A is the RMSD of an AC acceleration, v_0 is the AC speed in m/s, δ_φ is the RMSD of the azimuth angle of approximation on the course, and α is the coefficient.

For a regular flight $\alpha = \frac{1}{60}$; when maneuvering $\alpha = \frac{1}{20}$.

Any tracking system makes errors within $\pm\pi$ for the phase system and within half a stroke for the stroke system.

There is a suggestion to measure at the breaking stage the error between the reference signal and the received signal. Thus, tracking is carried out within the limits of $\pm\delta$, the value of the error, which is less than $\pm\pi$ or half the value of the stroke. It takes less time to synchronize. The greater is the signal-to-noise ratio, the less time it takes to coordinate. For initial conditions, synchronization is not spontaneous, but is relative to $\pm\delta$.

The final synchronization time of the T_c communication channel is one of the main factors limiting its functioning when transmitting navigation information. This occurs because errors in determination of the aircraft location, which are caused by the pulling into synchronism of a communication channel built on a digital basis, are proportional to this time.

$$\Delta = V_{AC} \cdot T_c, \tag{2.7}$$

where V_{AC} is the average speed of an aircraft.

The synchronization time of the communication channel T_c is found by adding of the time during which the search for the signal T_n occurs to the time during which the communication channel is entered into synchronism (establishment) T_{est}:

$$T_c = T_n + T_{est}. \tag{2.8}$$

During the transmission of digital data using the satellite channels AC-SC-GS or GS-SC-AC, the total synchronization time $T_{c\sum}$ is doubled, since the signal is received twice.

Let us assume:

$$T_{c\Sigma} \cong (2\ldots3)T_c. \tag{2.9}$$

Therefore, from (2.7) we have:

$$\Delta = V_{AC} \cdot T_{c\Sigma} \cong (2\ldots3)T_c V_{AC}. \tag{2.10}$$

The signal, defined as a signal with minimum frequency modulation (MFM), is the most promising, given the fact that it places the least severe conditions on the accuracy of synchronization of the receiving device. Therefore, the estimate of the time during which synchronization is performed must be made with respect to this signal. It is possible to consider the values of T_c calculated during this process as the lower limit for the time during which synchronization is performed.

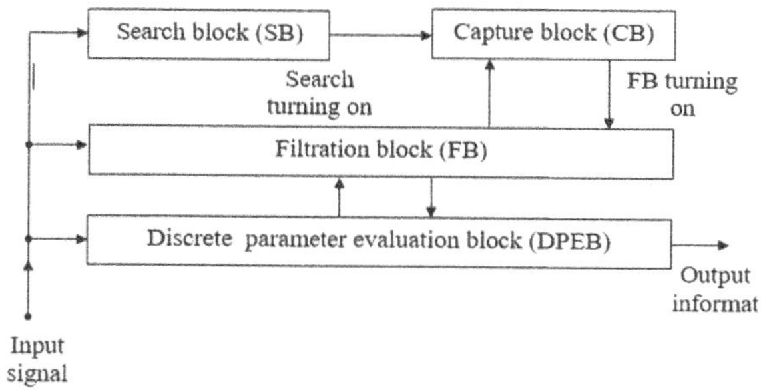

Fig. 2.10 The generalized block diagram of the receiving device of signals with MFM

To solve the problem of searching for signals, it is necessary to suggest the method for their recognition and detection by the value of the delay and the averaged frequency of the spectrum.

Let's consider, given in Fig. 2.10, the structure of the device receiving signals with MFM. This device uses the search procedure in two stages. During the first stage the signal enters the search block (SB), in which it is selected with the use of a synchronized filter. If the signal exceeds the threshold value in the SB, the block responsible for the capture (CB) is activated. The structure of this block includes the block that performs the filtering function (FB). This block performs the next, last synchronization step. It should be noted that the time delay and frequency values are introduced in the FB from the CB during the preliminary stage. The FB includes signal delay tracking (DTS) and automatic frequency control (AFC) systems. The synchronization system shown in Fig. 2.10, consisting of FB, SB and CB, carries out the formation of the reference signals received in the discrete parameter evaluation block (DPEB). Thus, it creates the opportunity to provide the flow of information. We can suggest the absence of multiple elements received with an error at the output of the DPEB as the criterion for receiver synchronization.

The most inertial links of the synchronization system include the following: signal delay tracking scheme (DTS), automatic frequency control (AFC) system. These systems are the components of the FB, while the time of transient processes in them T_{ycm}, makes the main contribution to the total synchronization time T_c.

Using the Markov model of the time delay τ of the signal and phase fluctuations φ, as well as the condition that the operation of the AFC and DTS schemes is independent, these systems are described by the following differential equation systems:

- For the AFC schemes [12, 16]:

$$\begin{cases} \dot{K}_{\varphi\varphi} = \frac{N_\varphi}{2} + K_{\varphi\varphi}^2 \overline{F}_{\varphi\varphi} + 2K_{\varphi\omega}, \\ \dot{K}_{\varphi\omega} = -\alpha\, K_{\varphi\omega} + K_{\omega\omega} + K_{\varphi\varphi} K_{\varphi\omega} \overline{F}_{\varphi\varphi}, \\ \dot{K}_{\omega\omega} = \frac{N_\omega}{2} - 2\alpha\, K_{\omega\omega} + K_{\omega\omega}^2 \overline{F}_{\varphi\varphi}; \end{cases} \tag{2.11}$$

- For the DTS scheme:

$$\begin{cases} \dot{K}_{\tau\tau} = \frac{N_\tau}{2} + K_{\tau\tau}^2 \overline{F}_{\tau\tau} + 2K_{\tau\Omega}, \\ \dot{K}_{\tau\Omega} = -\gamma\, K_{\tau\Omega} + K_{\tau\tau} + K_{\tau\tau} K_{\tau\Omega} \overline{F}_{\tau\tau}, \\ \dot{K}_{\Omega\Omega} = \frac{N_\Omega}{2} - 2\gamma\, K_{\Omega\Omega} + K_{\tau\Omega}^2 \overline{F}_{\tau\tau}, \end{cases} \tag{2.12}$$

where $N_\varphi, N_\omega, N_\tau$ and N_Ω are, respectively, one-sided spectral density of fluctuations of the frequency phase, the signal delay and the rate of its change; $K_{\varphi\varphi}$, $K_{\omega\omega}$, $K_{\tau\tau}$, $K_{\Omega\Omega}$ are the dispersions of the estimates of the corresponding parameters; $K_{\varphi\omega}$, $K_{\tau\Omega}$ are their coefficients of mutual correlation;

$$\overline{F}_{\varphi\varphi} \cong -\frac{A^2}{2N_0}\{t\,h(q) + t\,h(q/2)\}, \tag{2.13}$$

$$\overline{F}_{\tau\tau} \cong -\left(\frac{\pi}{2T}\right)^2 \frac{A^2}{2N_0}\{t\,h(q) + t\,h(q/2)\} \tag{2.14}$$

are the averaged values of functionals included in the equations of filtering of parameters that influence the slope in the DTS and AFC schemes of discriminatory characteristics; N_0 is the spectrum density for interference power; q is the signal-to-noise power ratio, A is the signal amplitude, γ and α are, respectively, the rate of change of the delay and the width of the spectrum of Doppler frequency fluctuations, and T is the stroke interval.

The duration of the transient process depends on two components: the signal-to-noise ratio, indicated by q, and the bandwidth of the tracking systems of DTS and AFC, which depend on the speed with which the delay γ changes and the width of the spectrum of signal frequency fluctuations α.

Based on the analysis of the solution of the system of equations for AFC (2.11), it can be concluded that in the case of the known in advance uniform distribution on the phase interval $[-\pi, \pi]$, the time during which the filtering error is set to the value that differs from its set-state value is not more than 10% at $\mu = \alpha T \cong 10^{-3}$, $q = 20$ $T_{ycm.} \cong (40 \ldots 50)\, T$.

In the same way, the time of establishment is determined for DTS. Based on the use of interconnectedness of parameters φ and τ at the complex use of AFC and DTS, the establishment time doubles. It is [20]:

$$T_{est} \cong (80 \ldots 100)\, T. \tag{2.15}$$

Based on the use of time in the calculations, during which the search for a signal equal to $T_n \cong 0{,}2\, T_{est}$, occurs, from (2.15) we have:

$$T_c \cong (100 \ldots 120)\, T, \tag{2.16}$$

which corresponds to the conditions imposed on the original synchronization.

Therefore, assuming that $T_c \cong 120$ T, from (2.16) we have:

$$T_{c\Sigma(max)} \cong 3\,T_c \cong 360\,T. \tag{2.17}$$

Therefore, from the expression (2.17) we have the following expression for the positional error:

$$\Delta \cong 360\,T\;V_{AC}. \tag{2.18}$$

Taking into account the connection of the stroke interval T with the speed at which data is transmitted ϑ by the expression

$$\vartheta = 1/T, \tag{2.19}$$

and measuring ϑ in bits/sec, V_{BC} in km/h and Δ in meters, we obtain from (2.18) the last relation:

$$\Delta_{(M)} = 100\,V_{AC}\,[\text{km/hr}]/\vartheta[\text{bit/sec}]. \tag{2.20}$$

The final positioning accuracy of an aircraft based on the available information on positional errors, which are a consequence of the entering of the communication channel into synchronism, is determined by the expression

$$\sigma_{рез.} = \sqrt{\sigma^2 + \Delta^2}\,, \tag{2.21}$$

where σ is the positioning error on an aircraft board based on the use of the navigation system.

Since in GPS SRNS, the root mean square error (RMSE) equals to $\sigma = 50.6$ m under the condition of using the C/A open code, which has reduced accuracy in nominal mode with the satisfactory geometric factor of 2, from (2.21) we find that data transfer rate $\vartheta = 1200$ bps, for which, as noted, the positional error is $\Delta =$ 70 m. The resulting positioning error is 86.4 m, which is by 1.71 times exceeds the original positioning error of 50.6 m. At data transfer rates $\vartheta = 600$ bps and $\vartheta = 300$ bps for which the positioning error is, respectively, 140 m and 280 m, the resulting positioning error is 148.9 m and 284.5 m, which is by 2.9 and 5.6 times exceeds the original positioning error, that is, it is almost determined by the positional error. In this regard, the problem of reducing the synchronization time and, accordingly, the positional error is actual.

The dynamic characteristics of a communication channel synchronization system can be significantly improved by optimizing the performance of the AFC and DTS systems, which, as noted, determine the speed of the synchronization system as a whole.

$$g(\varphi) = sign\,\varphi,\;\; |\varphi| \le \pi \tag{2.22}$$

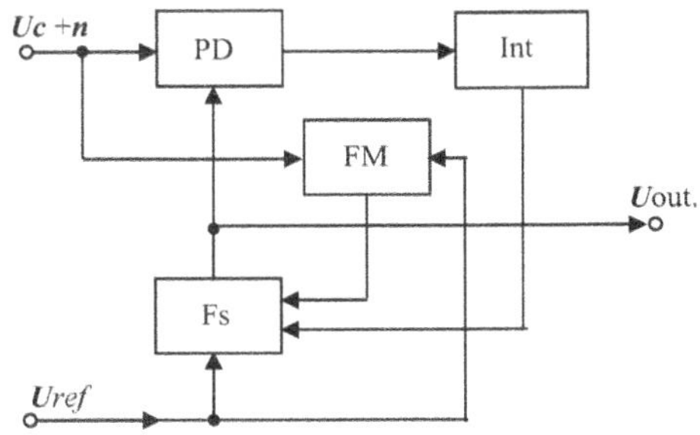

Fig. 2.11 The block diagram of the tracking system with improved dynamic characteristics

The AFC system with first-order astaticism and relay performance of the phase discriminator (PD) is optimal on speed-of-response in the absence of interference and is close to optimal at its low level. The additional reduction of the transient process time in the system can be achieved if at its turning on the signal phase tracking ring opens and the phase shifter in the reference signal circuit is set to the position corresponding to the instantaneous value of the phase of the signal and interference mix during a one-time reading, which pre-synchronizes the system.

The block diagram of the tracking system with improved dynamic characteristics, which can be used as the AFC system, is shown in Fig. 2.11.

Where $Uc + n$ is a mixture of signal and interference, *Uref* is reference signal, *Uout* is output signal, PD is phase discriminator, Int is integrator, Fs is phase shifter and FM is phase meter [21].

Physically, the gain in transient process time due to such an improvement in the AFC system is due to the fact that if in a conventional system initial displacement $\Delta\varphi$ is distributed according to the uniform law [22–24]

$$W(\Delta\varphi) = 1/2\pi, \ |\Delta\varphi| \leq \pi, \tag{2.23}$$

then in the considered system it is grouped near zero and is determined by the phase distribution of the signal and interference mixture.

In estimating the gain, for simplicity, we will assume that the phase of the signal varies under the linear law

$$\psi(t) = \psi_0 + \omega_g t, \tag{2.24}$$

where ψ_0 and ω_g are, respectively, the initial phase and the Doppler frequency shift.

The transition process time in the system in the absence of interference can be determined from the following differential equation describing its behavior

$$d\psi/dt = kg(\varphi), \quad \varphi = \psi - \psi^*, \tag{2.25}$$

where k is the open-loop transmission coefficient, which determines the effective bandwidth of AFC ($\Delta\omega = k / 2$), ψ^* is the estimated value of the signal phase.

Separating in (2.25) the variables and taking into account (2.24), we have

$$dt = d\varphi / k[c - g(\varphi)], \tag{2.26}$$

where $c = \omega_g / k$. Integrating (2.24) under boundary conditions: $\varphi = \Delta\varphi$ at $t = 0$, $\varphi = \varphi_0$ at $t = T_{est.}$, we obtain the expression for the time of the transition process (establishment)

$$T_{ycm.} = k^{-1} \int_{\varphi_0}^{\Delta\varphi} [g(\varphi) - c]^{-1} dk. \tag{2.27}$$

In the presence of interference $T_{est.}$ is a random value. The average transition time $T_{est.}$ in a real system with interference coincides with the time of the transition process in an equivalent system without interference, in which the coordinate deviations coincide with the mathematical expectations of the deviations of the coordinates of the real system. Hence, for the average time of the transition process with the initial displacement $\Delta\varphi$ taking into account (2.27) we have

$$\overline{T_{ycm.}(\Delta\varphi)} = k^{-1} \int_{\varphi_0}^{\Delta\varphi} \left[\overline{g(\varphi)} - c\right]^{-1} d\varphi, \tag{2.28}$$

where $\overline{g(\varphi)}$ is the averaged characteristic of PD that is determined by taking into account the cyclic phase from the expression

$$\overline{g(\varphi)} == \int_{-\pi}^{\pi} g(\upsilon - \psi) W(\vartheta / \psi) d\upsilon = \int_{-\pi}^{\pi} g(\upsilon) W(\upsilon / \varphi) d\vartheta, \tag{2.29}$$

where $W(\upsilon / \varphi)$ is the density of the probability of the phase of mixture of the signal and interference, which coincides (at $\varphi = 0$) with the distribution $\Delta\varphi$ in the system with preliminary synchronization [25].

Thus, averaging $\overline{T_{ycm.}(\Delta\varphi)}$ additionally on $\Delta\varphi$ and taking into account (2.28), we obtain

$$\overline{T_{ycm.}} = k^{-1} \int_{-\pi}^{\pi} W(\Delta\varphi) \int_{\varphi_0}^{\Delta\varphi} \left[\overline{g(\varphi)} - c\right]^{-1} d\varphi \, d(\Delta\varphi). \tag{2.30}$$

For the case of the absence of preliminary synchronization, substituting in (2.30) $W(\Delta\varphi)$ from (2.23), we obtain

$$\overline{T_{ycm.o}} = (2\pi k)^{-1} \int_{-\pi}^{\pi} \int_{\varphi_0}^{\Delta\varphi} \left[\overline{g(\varphi)} - c\right] d\varphi \, d(\Delta\varphi). \tag{2.31}$$

The gain in the mean time of the transition process due to the preliminary synchronization is

$$M = \overline{T_{ycm.o}} / \overline{T_{ycm.}}$$
$$= \int_{-\pi}^{\pi} \int_{\varphi_0}^{\Delta\varphi} \left[\overline{g(\varphi)} - c\right]^{-1} d\varphi \, d(\Delta\varphi) / 2\pi \int_{-\pi}^{\pi} W(\Delta\varphi) \int_{\varphi_0}^{\Delta\varphi} \left[\overline{g(\varphi)} - c\right]^{-1} d\varphi \, d(\Delta\varphi) \cdot \tag{2.32}$$

It is reasonable to take the final displacement φ_0 in (2.32) equal to the instrumental error of the system.

We present an analysis of the speed-of-response of the improved system under the conditions of normal interference, whose probability of density of the phase of the mixture with the signal is

$$W(\upsilon/\varphi) = \frac{1}{2\pi} + \frac{s}{\sqrt{2\pi}}(\upsilon - \varphi) F[s \, \mathbf{cos}(\upsilon - \varphi)] \times \exp\left[-\frac{s^2}{2} \mathbf{sin}^2(\upsilon - \varphi)\right], \quad |\upsilon - \varphi| \le \pi, \tag{2.33}$$

where $s = A_c / \sigma_n$ is the ratio of the amplitude of the signal to the RMS value of the interference, $F(\cdot)$ is the Laplace function [26].

Substitution of $g(\varphi)$ from (2.22) and $W(\vartheta / \varphi)$ from (2.33) to (2.29) gives the following expression for the averaged PD characteristic

$$\overline{g(\varphi)} = 2F(s \, \mathbf{sin} \, \varphi) - 1, \quad |\varphi| \le \pi. \tag{2.34}$$

Substituting further $W(\Delta\varphi) = W(\upsilon/\varphi)\Big|_{\vartheta = \Delta\varphi}^{\varphi = 0}$ from (2.33) and $\overline{g(\varphi)}$ from (2.34) into (2.32), we obtain the gain M in the value of the mean time of the transition process due to the preliminary synchronization. The results of the calculation of the dependence of M from the signal-to-noise ratio (on power) $q = s^2 / 2$ for two values of the parameter c: $c = 0$ and $c = 0.5$, corresponding to the absence of the Doppler frequency shift and the value of the Doppler shift ω_g, equal to the effective bandwidth of the system $\Delta\omega$ ($c = \omega_g / k = \omega_g / 2 \, \Delta\omega | \omega_g = \Delta\omega = 0.5$), are shown in Fig. 2.12. At the same time, we take $\varphi_0 = 1^\circ$.

As you can see, the gain in the mean transition time due to pre-synchronization is significant and increases with increase in the signal-to-noise ratio.

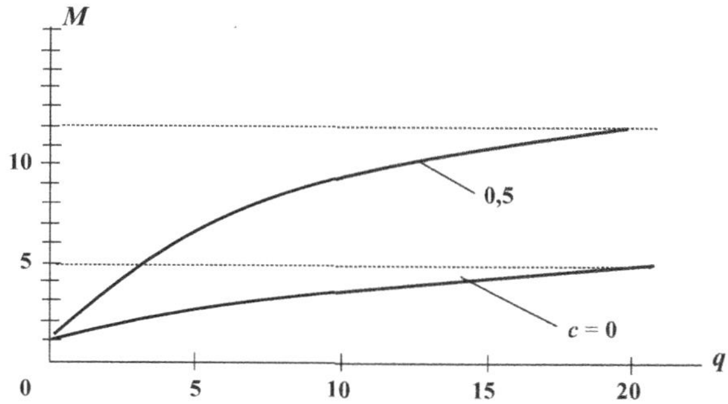

Fig. 2.12 Gain in the mean time of the transition process

So, with the value $q = 20$, taken above when assessing the positional errors of an aircraft positioning connected with the entry into synchronism of a digital communication channel, the gain when changing the Doppler frequency shift within $0 \ldots \Delta\omega$ is $M = 5 \ldots 12$.

Similarly, the dynamic characteristics of the DTS can be improved. Assuming the gain in synchronization time $M \geq 5$, taking into account the fact that the positional error Δ is proportional to the total synchronization time $T_{c\sum}$, determined mainly by $T_{est.}$ (see (2.8) –(2.10)), it can be considered that the positional error decreases approximately in the same ratio. At the same time, the position error value obtained above $\Delta = 140$ m for the medium-speed data transmission with $\vartheta = 600$ bps decreases to the value $\Delta_1 \cong \Delta / M \leq 28$ m, and the resulting aircraft positioning error at the accepted above standard deviation in the GPS SRNS (the C/A code) $\sigma = 50.6$ мm to the value $\sigma_{pe3.1} = \sqrt{\sigma^2 + \Delta_1^2} \leq 57.8$ м.

As you can see, the proposed method of improvement of the dynamic characteristics of the AFC and DTS systems makes it possible to eliminate virtually the influence of positional errors on the resulting aircraft positioning error on the SRNS.

References

1. Apollonov AA (2010) Improving the accuracy and reliability of user equipment in high latitudes when using pseudosatellites. Sci Bull MSTUCA 169
2. Amirkhanov AV (2017) The issues of pairing of the system of instrument-technological modelling with the library of optimization algorithms. In: Amirkhanov AV, Glushko AA, Gladkikh AA, Makarchuk VV, Shahnov VA (eds) Nanoindustry, 74, pp 389–394
3. Gnedenko BV (1965) Mathematical methods in the theory of reliability. In: Gnedenko BV, Belyaev YK, Solovyov AD, Znanye M (eds) Publishing house Nauka, Main editorial board of physical and mathematical literature, p 524
4. Zatuchny DA (2010) Identification of false information on the voice of the pilot. In: Proceedings of the international symposium Reliability and quality. vol 1, pp 429–430
5. Pavlov IV (1972) Calculation of confidence limits for the functions of many unknown parameters. Tech Cybern 2

6. Pavlov IV (1979) Confidential assessment of reliability and efficiency of complex systems based on the results of reliability tests. In: Barsky YD, M (eds) Znanye Issues of experimental evaluation of reliability indicators: the collection of articles, p 90
7. Pavlov IV (1976) Interval assessment of the reliability of the system according to the reliability of its components. Reliab Quality Control 10
8. Pavlov IV (2017) Evaluation of the reliability of the system with redundancy based on the results of tests of its elements. Autom Remote Control (3):149–158
9. Severtsev NA (2014) Security and protection of complex systems. In: - M (ed) Dorodnitsyn Computing Centre of RAS, p 238
10. Severtsev NA (2013) Security and fault tolerance of dynamic systems. In: - M (ed) Culture and technology, p 412
11. Severtsev NA, Beckov AV (2009) System analysis of the theory of security. In: M (ed) Theis, p 452
12. Teskin OI (1979) Exact confidence limits for the reliability of reduced systems for trouble-free testing. Techn Cybern (4)
13. Ushakov IA (1975) Reliability of complex information control systems. In: Ushakov IA, Gadasin VI, - M (1975) Soviet radio, p 192
14. Ushakov IA (1970) Engineering methods for calculation of reliability. In: – M (ed) Znanye, vol 4
15. Yarlykov M.S. Fight use and effectiveness of aviation radio-electronic complexes / M.S. Yarlykov, A.S. Bogachev, M.A. Mironov // Ed. Yarlykov M. S. – M: VVIA named after the prof. N. E. Zhukovsky, 1990. - 250 pages
16. Yarlykov MS (1993) Markov theory of estimation of random processes. In: Yarlykov MS, Mironov MA, – M (eds) Radio and communication, p 461
17. Yarlykov MS (2004) Markov theory of estimation in radio engineering. In: Yarlykov MS, Zavalishin OI, Akinin AL, Yarlykov MS, – M (eds) Radio engineering, p 503
18. Lloyd DK, Lipov M (1962) The reliability: management methods and mathematics. Prentice-Hall. Inc., Englewood Cliffs, New Jersey
19. Ushakov IA (1999) Handbook of reliability engineering. Wiley, New York
20. Zatuchny DA, The assessment of the functional reliability of the combined means of communication of air traffic management systems and navigation support of aircraft based on satellite radio navigation systems: Thesis for the degree of candidate of Technical Sciences. 05.22.13 / Zatuchny Dmitry Aleksandrovich
21. Buslenko NP, Shreyder YA (1961) Statistical test method (Monte Carlo) and its implementation in digital machines. In: - M (ed) Science
22. Pichugin SM (2015) Improving the accuracy and reliability of the complex navigation tools for flight assurance with the use of response on the solution of the navigation task. Thesis for the degree of candidate of Technical Sciences, Moscow, MSTUCA
23. Proshin MV (2001) Improving the efficiency of navigation support of aircraft by integrating satellite navigation systems with other navigation and radio communications facilities: Thesis for the degree of candidate of Technical Sciences. – M.: MSTUCA, p 149
24. Baranov EV (2008) Improving the reliability of assessing the accuracy of navigation definitions for satellite navigation systems by the methods of semi-natural modelling Thesis for the degree of candidate of Technical Sciences. Speciality 05.22.13—Navigation and air traffic control. In: Baranov EV, – M (eds) MSTUCA, p 129
25. Slepchenko AP (2011) Optimization of the conditions of the navigation session of the objects of aviation transport system: Thesis for the degree of candidate of Technical Sciences. 05.13.01. Slepchenko A.P.; [Place of defence: Irkutsk State Transport University]. - Irkutsk, p 157
26. Filin AD, Shatrakov YG, Yakovlev VT, Yushkov AV (2017) The impact of reliability of flight support subsystems on the likelihood of occurrence of flight accidents. Sci Bull Military-Ind Complex Russia (3):68–74

Chapter 3
Methods to Improve Noise Immunity, Integrity and Continuity of Service of a Navigation System at the Performing Stage of Terminal Procedures and Precision Approach

3.1 Improving the Continuity Characteristics of the Complex Navigation System in an Airfield Area and During a Precision Approach Based on the Use of Two-Constellation Functional Extensions to Ground-Based GNSS (Multi-constellation GBAS)

About 120 LMCS (GBAS) which meet the requirements of SARPs for the GBAS ground-based augmentation systems currently operate in the Russian Federation. These stations provide highly accurate navigation of aircraft in the code differential mode (DM) and the approach of the CATI category (GASTC) with the use of civilian signals of the L1 GLONASS and GPS range.

However, the provision of a landing approach and landing under the CATII, CATIII (GASTD) categories with the use of civilian signals of the L1 range only leads to the necessity of using remote monitoring receivers in the GBAS (LMCS) that are designed to provide control of the spatial gradient of the ionospheric anomalies required to perform the set characteristics of integrity. This design decision is expensive and makes it very difficult to deploy LMCS at a GAST D level. The solution to this problem can be simplified by comparing the two-frequency estimates of ionospheric delays by basic receivers (BR) of LMCS and aircraft receivers with the use of civil signals of the L1, L2 (L3, L5) ranges [1].

On the other hand, the use of Galileo and Beidou (Great Bear) signals, along with the already used GLONASS and GPS GNSS signals, will significantly improve the availability and continuity of GBAS systems in airfields areas, in which part of the sky is shaded by closely located mountain ranges. The specified reasons determine the relevance of the development of multi-frequency multi-system (GLONASS/GPS/Galileo/Beidou) ground-based MF/MC GBAS (multi-frequency/multi-constellation GBAS) functional systems of the CAT II/III (category II/III). The creation of such systems will allow:

O. I. Zavalishin et al., *Modern Requirements for Noise Immunity Aircraft Navigation Equipment*, Springer Aerospace Technology,
https://doi.org/10.1007/978-981-16-0073-9_3

- to solve fundamentally the problem of ensuring the integrity of GBAS in the presence of ionospheric anomalies (ionospheric storms) due to the transmission to users of two-frequency ionospheric delay estimates of the LMCS BR;
- to improve the quality of the solution of the navigation task in the DM due to the use of smoothing of the measured pseudo ranges by taking into account the increments of the linear combination of complete phases, free from the influence of the ionosphere;
- to provide high level of GBAS service continuity in airfields areas, in which part of the sky is shaded by closely located mountain ranges, by increasing the number of simultaneously observed NS using all available GNSS.

The results of measurements of coordinates by GPS + GLONASS + Galileo are confirmed by the criterion for ensuring the landing accuracy of the CATI by almost 100% continuity of the characteristics of the geometric factor. Due to the fact that modern ICAO standards consider the use of only two groups of GPS and GLONASS, further calculation was made on the two-constellation system.

The risk of loss of continuity of service of the GBAS system is defined as the probability of the coordinate estimation error beyond the maximum permissible values for the period of operation, provided that at the start of the operation the coordinate estimation errors were within the tolerance [2].

To calculate the continuity characteristics of the GBAS system, we will use the method of evaluation of the effectiveness of complex technical systems (CTS). The probability of the system performance of this task is usually taken as a general indicator of the effectiveness of the CTS. A system of any class can be characterized by some finite set of incompatible states $\{H_i\}$, $i = \overline{1, M}$, where M is the total number of these states. Each of the CTS states H_i at the time point t is characterized by the unconditional probability of this state $P_i(t)$ and it can be associated with the value of some function $\omega_i(t)$, determining the quality of the solution by the CTS of the task under consideration. The function $\omega_i(t)$ is called the conditional indicator of efficiency, since it corresponds to a certain state of the CTS *Hi*.

The assembly average of the function $\omega_i(t)$ is taken as an indicator of the effectiveness of the CTS, which is equal to [3, 4]:

$$\omega(t) = \mathbf{P}^T(t)\boldsymbol{\omega}_y(t) = \sum_{i=1}^{M} P_i(t)\omega_i(t), \tag{3.1}$$

where $\omega(t)$ is the indicator of the effectiveness of the CTS (probability of the system performing the task); $\mathbf{P}(t) = [P_i(t)]$, $\boldsymbol{\omega}_y(t) = [\omega_i(t)]$ are the vectors columns of unconditional probability of the CTS and the conditional performance indicators corresponding to them.

When solving a navigation task in the DM, it is advisable to use the probability that at the considered (calculated) point in time an aircraft is in the area of safety, in which the error of coordinates do not exceed the maximum permissible values as a conditional indicator of the effectiveness of the GBAS system. The conditional performance indicator in this case can be represented as the conditional probability [5].

$$\omega_i(t) = \int_{-h_B}^{h_B} \int_{-h_L}^{h_L - h_H} \int_{H_H}^{h_H} p(t, \varepsilon_B, \varepsilon_L, \varepsilon_H | H_i) d\varepsilon_B d\varepsilon_L d\varepsilon_H, \quad (3.2)$$

where h_B, h_L, h_H are the maximum allowable errors in the determination of coordinates.

When performing landing approach and landing operations using GBAS, the conventional performance indicators (3.2) depend on the number of navigation satellites (NS) used to solve the navigation task (NT). In turn, this number of NS is determined both by the properties of the GNSS system(s) and the results of the operation of the integrity control (IC) algorithms of the LMCS [6].

In normal mode, the number of observed NS of the GPS system is generally in the range from 6 to 14. The most likely values are 7–11. Less often, 6, 12, 13, 14.5 or less, as well as more than 14 NS are practically not observed. Similarly, for the GLONASS system, 6–11 NS are mainly observed. For two GPS and GLONASS systems, as a rule, more than 14 NS are simultaneously observed, most likely 17–20 NS. To estimate the RMSD of the errors of the NT solution in the DM, depending on the number of observed NS, their experimental implementations of 100,000 half-second epochs were used [7].

As an example, Figs. 3.1, 3.2 and 3.3 show the implementation of the errors of estimating the coordinates in the DM when solving the NT on 5, 6 and 7 NS of the GPS system, obtained for the same raw data record with artificial limitation of the number of NS. The implementations of estimation errors for the same record with the use of all visible and non-rejected NS integrity control (IC) algorithms are shown in Fig. 3.4.

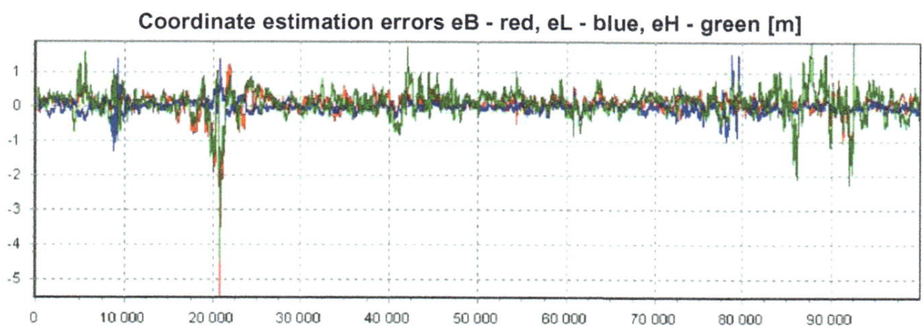

Fig. 3.1 Realization of errors in the estimation of coordinates in the DM when solving the NT for 5 NS of the GPS system

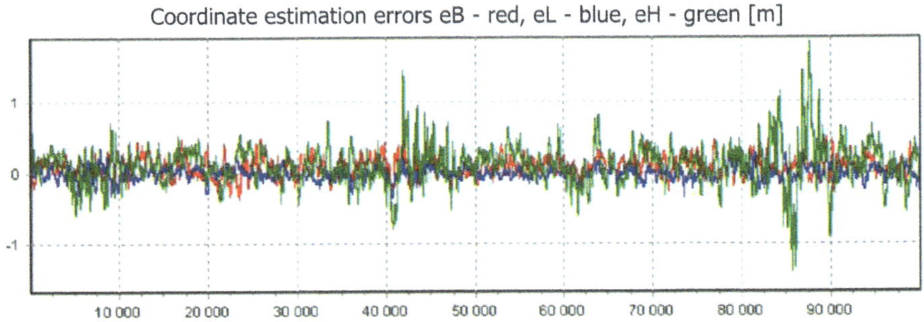

Fig. 3.2 Realization of errors in the estimation of coordinates in the DM when solving the NT for 6 NS of the GPS system

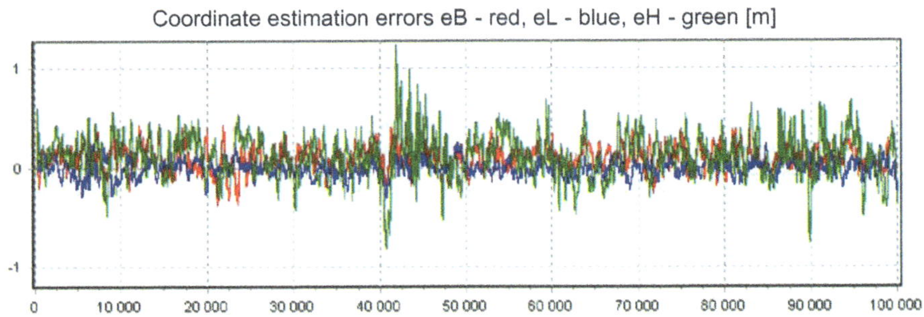

Fig. 3.3 Realization of errors in the estimation of coordinates in the DM when solving the NT for 7 NS of the GPS system

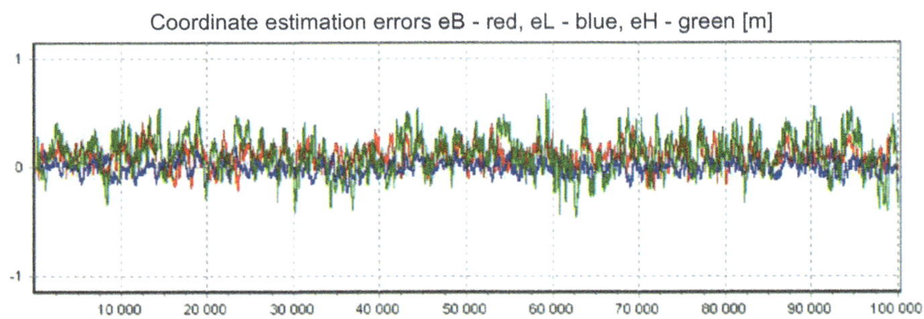

Fig. 3.4 Realization of estimation errors for the same record when using all visible and non-rejected by IC algorithms NS

The total number of these NS used to solve the NT with 11 rejections by all IC algorithms is shown in Fig. 3.5.

To calculate the unconditional probability of the states of a CTS, the possible transitions from state to state are usually described by a Markov chain.

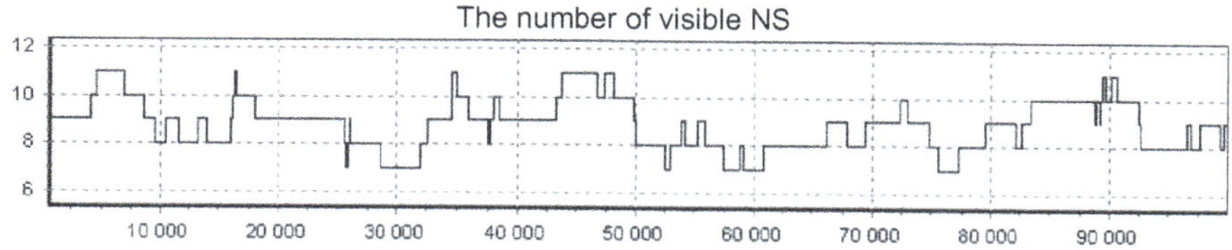

Fig. 3.5 The total number of NS used to solve the NT taking into account 11 rejections

However, as applied to the GBAS service continuity assessment problem, this approach encounters difficulties connected with determination of the elements of the corresponding transition probability matrix. Therefore, the approach based on the experimental evaluation of empirical probabilities of states, i.e., the probabilities of using for the NT solution of 4, 5, 6, 7, 8 and more NS by taking into account rejections by IC algorithms is further used.

To estimate these probabilities, we used realizations of the number of NS used to solve NT on the GPS + GLONASS, GPS and GLONASS systems during the month of LMCS operation taking into account rejections by IC algorithms (such as shown in Fig. 3.5). Every day the probabilities of using a given number of NS for solving the NT were determined based on these implementations by counting the corresponding number of epochs. The examples of the obtained histograms are shown in Fig. 3.6. (in yellow for the GLONASS system, blue for the GPS and red for the GPS + GLONASS).

When installing LMCS in some airports (e.g., in Adler), a smaller number of NS may be observed due to the shading of a part of the celestial sphere with local objects

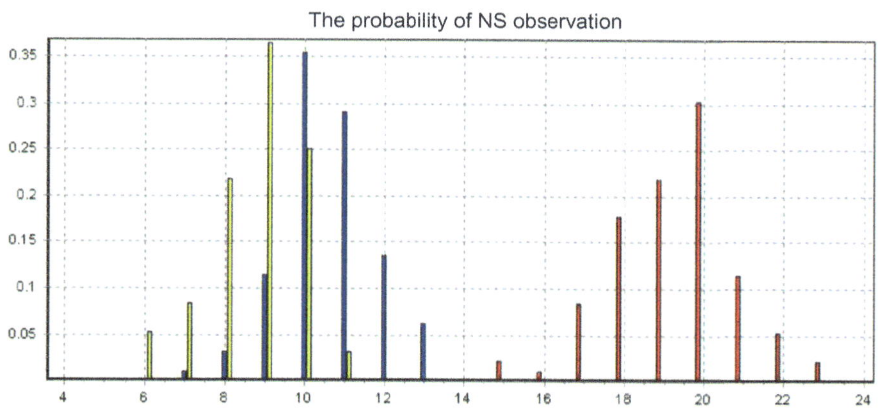

Fig. 3.6 Implementations of the number of NS used to solve the NT for GPS + GLONASS, GPS and GLONASS

(mountains). When calculating the GBAS service continuity characteristics, this fact can be taken into account by increasing the probability of observing a smaller number of NS and a corresponding decrease in the probability of observing a larger number of NS [2, 8].

For the stationary mode, when the unconditional probability of a state H_i does not depend on time, i.e., $P_i(t_{k+1}) = P_i(t_k) = P_i$, the efficiency (3.1) is determined by the ratio

$$\omega = \sum_{i=1}^{M} P_i \omega_i \tag{3.3}$$

Thus, according to (3.3), the probabilities of solving the NT in the DM in the normal mode (the values in brackets are shown for a limited constellation) are [9]:

– For the GPS system:

CATII $1 - 10^{-11}(1 - 1.64 \cdot 10^{-7})$;
CATIII $1 - 1.041 \cdot 10^{-7}(1 - 1.81 \cdot 10^{-6})$;

– For the GLONASS system:

CATII $1 - 2 \cdot 10^{-11}(1 - 1.645 \cdot 10^{-7})$;
CATIII $1 - 2.082 \cdot 10^{-6}(1 - 6.7 \cdot 10^{-6})$;

– At the joint use of GPS and GLONASS:

CATII $1.0(1 - 5 \cdot 10^{-12})$;
CATIII $1 - 10^{-9}(1 - 5.301 \cdot 10^{-8})$.

The required service continuity characteristics are:

CATII $1 - 4 \cdot 10^{-6}/15 = 1 - 2.6 \cdot 10^{-7}$;
CATIII $1 - 2 \cdot 10^{-6}/30 = 1 - 6 \cdot 10^{-8}$.

This, in particular, implies that with a limited constellation the use of only one system sometimes does not allow to meet the requirements of SARPs to service continuity. The joint use of the two systems satisfies the requirements for all categories of landing approach and landing under the normal conditions of operation of these systems. However, under the influence of interference, organized or caused by the operation of other radio equipment (RE) in the LMCS or AC zone, there are losses in tracking of GNSS signals.

The continuity characteristics of the complex GNSS system and an angular optical system are largely determined by the structure of this system. For definiteness, let us assume that using the optical system, aircraft coordinates are estimated with a set accuracy, even if with a lower accuracy compared to the GNSS system.

The simplest, but rather effective way of integration, is the inclusion of estimates of the optical system in the observation vector of LSM algorithms. In this case, the structure of the NT solution algorithms based on GNSS signals remains unchanged, but the properties of the system change drastically:

- When tracking signals of less than four satellites, the NT solution with standard LSM algorithms is impossible in principle, while the extended observation vector allows to form LSM estimates even with the loss of tracking of all satellites;
- The dependence of the accuracy of estimation of the coordinates from the geometric factor and noise conditions decreases significantly.

To calculate the continuity characteristics of the complex system service, the same methodology is used as in the clause 3.1.1: for the stationary mode, when the unconditional probability of the state H_i does not depend on time, i.e., $P_i(t_{k+1}) = P_i(t_k) = P_i$, the efficiency (3.1) is determined by the correlation (3.3). The values of the probabilities P_i remain unchanged, and only the conditional probability changes ω_i. The probabilities of solving the NT in the DM under the conditions of a difficult interfering environment are equal to:

- For the GPS system:

  CATII $1 - 3.67 \cdot 10^{-7}$;
  CATIII $1 - 1.43 \cdot 10^{-6}$;

- For the complex system

  CATII $1 - 2.7 \cdot 10^{-9}$;
  CATIII $1 - 3.4 \cdot 10^{-8}$.

Thus, the use of complex coordinate estimation algorithms according to GNSS and goniometric optical system increases the characteristics of service continuity.

Let us consider the possibility (within the framework of the existing ICAO format) of data transmission in the VDB channel for three groups and two frequencies.

Table 3.1 shows the calculation of the "budget" of the VDB channel, from which it is clear that the transmission line of real GBAS (e.g., LMCS-A-2000) provides the transmission of navigation landing information not only by GPS/GLONASS, but also by Galileo, including second civilian frequencies and CASTD [10].

To assess the dependence of the accuracy of the coordinates on the location of the SNS antennas, we present the results of satellite signal measurements performed at three different airports [4, 11].

Operating mode: autonomous.

The Figs. 3.7 and 3.8 show the errors in determination of the coordinates and the number of visible satellites for the Samara airport [5, 9].

The coordinates are 53°30′ n.l. 50°09′ e.l.

The Figs. 3.9, 3.10 and 3.11 show the errors of determination of the coordinates in the GPS system and the GLONASS system, as well as the number of satellites used for the Gorno-Altaisk airport.

The coordinates are 51°58′ n.l. 85°50′ e.l.

Table 3.1 Calculation of the "budget" of the line of navigation landing information

Satellite systems	CAT I		CAT II/ III		CAT II/ III MF**	
	Slots are occup	Bytes available not less than	Slots are occup	Bytes available not less than	Slots are occup	Bytes available not less than
GPS	2	1266	3	1055	3	1055
					4	844
GPS + GLONASS	3	1055	5	633	5	633
					7	211
GPS + GLONASS + GALILEO	3[a]	1055	5[a]	633	5	633
	4[a]	844	6[a]	422	8	–

[a]Dimension of messages are given for different standardization cases.

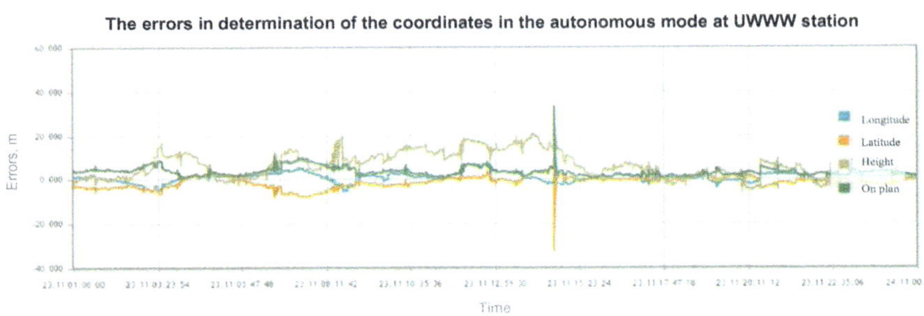

Fig. 3.7 Coordinate errors (GLONASS + GPS)

The Figs. 3.12, 3.13, 3.14 and 3.15 show the errors of determination of the coordinates in the GPS system, the GLONASS system, the GPS + GLONASS system and the number of visible satellites for the polar airport—Novolazarevskaya station [12].

The coordinates are 70°46′ s.l. 11°49′ e.l.

The analysis of the obtained statistical characteristics of measurements of the accuracy of coordinates clearly shows that the accuracy of coordinates decreases with a decrease in the number of visible satellites (up to 50%, Gorno-Altaisk), as well as with the impairment in the satellite geometry (up to 20%, Antarctica).

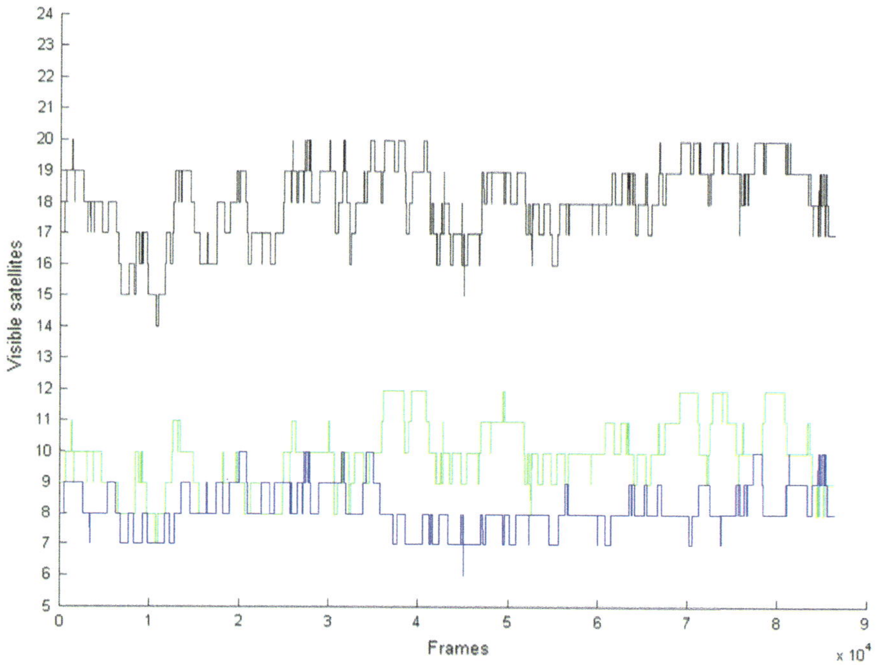

Fig. 3.8 The graph of the number of satellites used in solving the navigation task (black: GLONASS + GPS; green: GPS; blue: GLONASS)

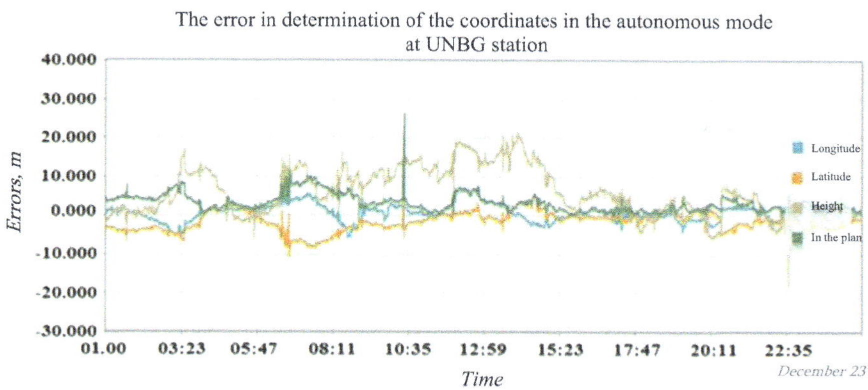

Fig. 3.9 Coordinate errors (GPS)

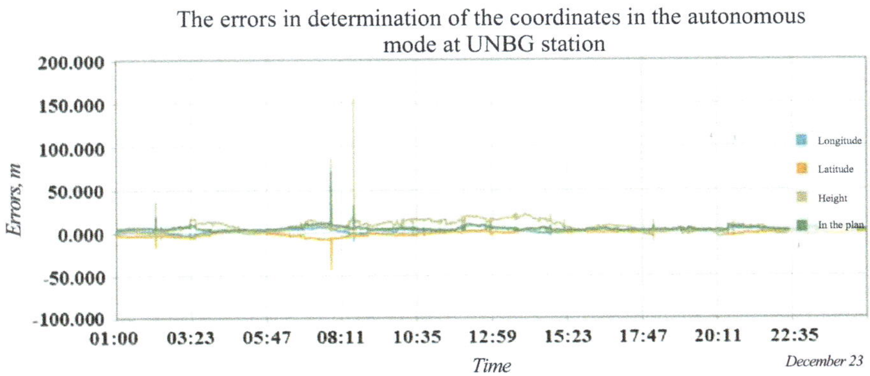

Fig. 3.10 Coordinate errors (GLONASS)

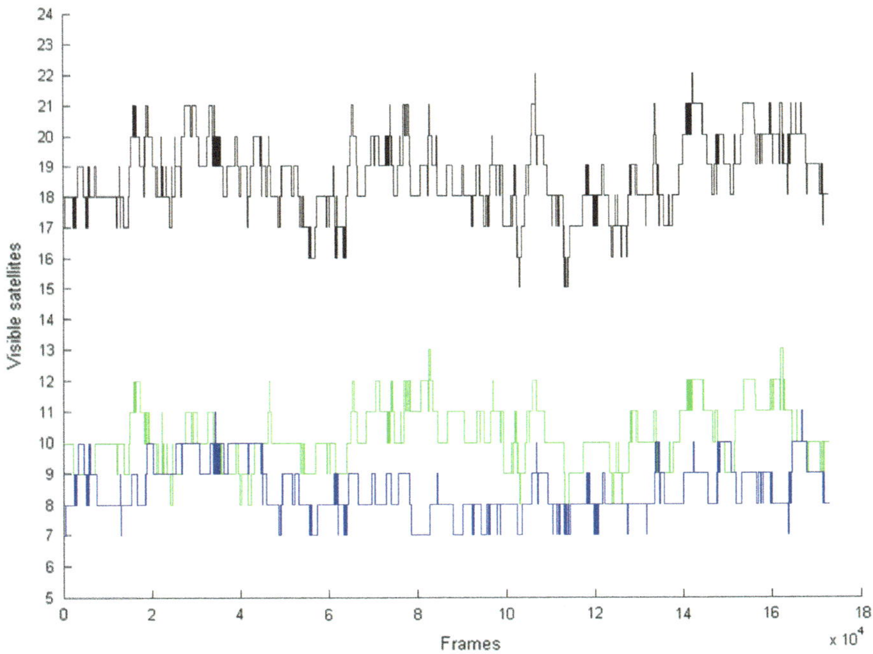

Fig. 3.11 The number of used satellites (black: GLONASS + GPS; green: GPS; blue: GLONASS)

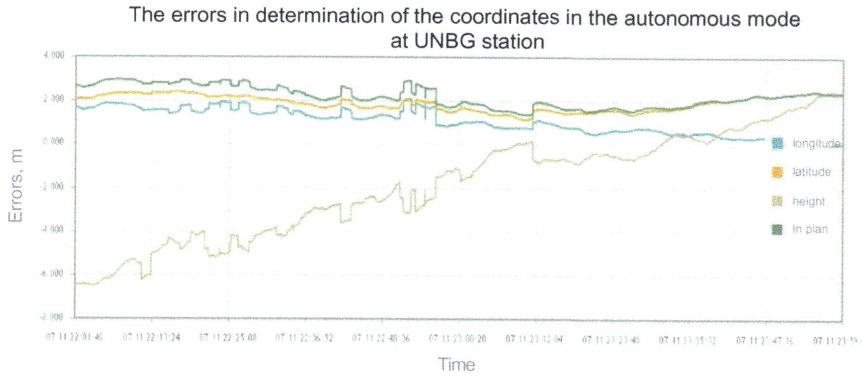

Fig. 3.12 GPS coordinate errors

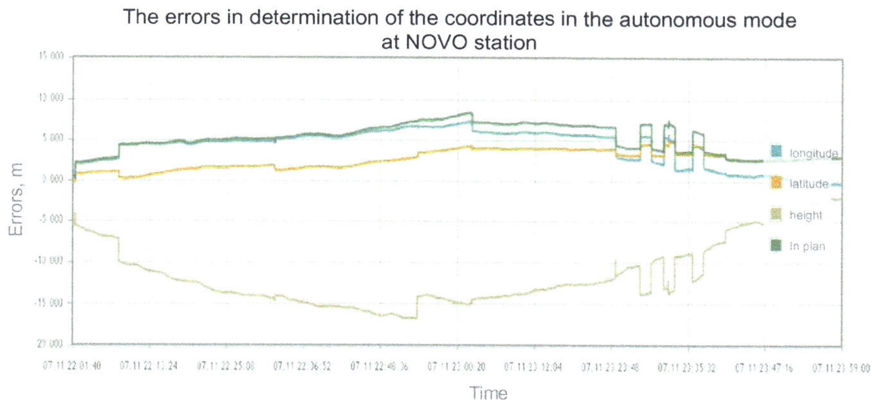

Fig. 3.13 GLONASS coordinate errors

3.2 Improving the Noise Immunity of the Integrated Navigation System in an Airfield Area and at Precision Landing Approach Based on the Use of Two-Constellation Functional Extensions to the Ground-Based GNSS (Multi-constellation GBAS)

The integrity at the level required for information support of landing approach procedures of the ICAO category II/III planting in the SLS (GBAS) is provided by the integrity control algorithm that includes three levels [13–15]:

1. Quality control of pseudo range and full phase measurements (MQM), navigation satellite signals (SQM), and ephemerical information (DQM) at the level of each basic receiver, up to the formation of differential corrections. This makes it

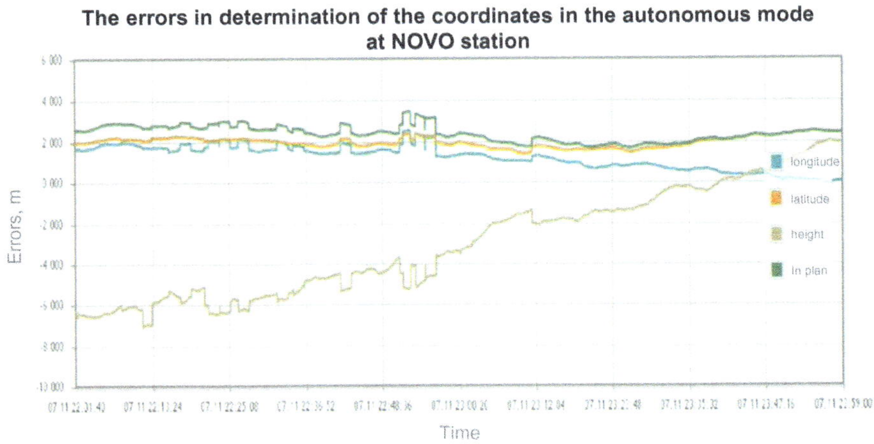

Fig. 3.14 GLONASS + GPS coordinate errors

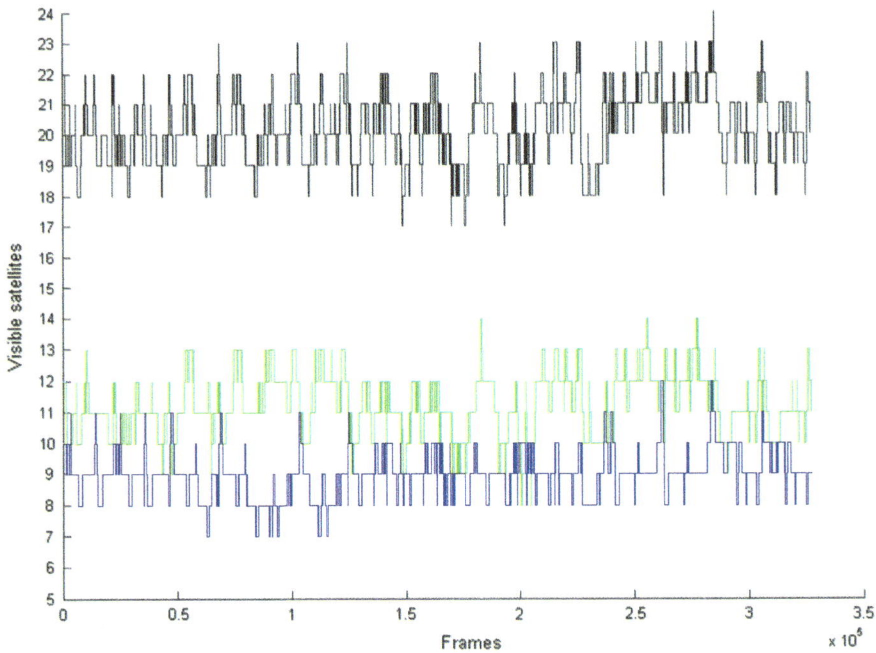

Fig. 3.15 The number of visible satellites (black: GLONASS + GPS; green: GPS; blue: GLONASS)

possible to detect the presence of distortions in the pseudo random sequence of the signal from a navigation satellite, the anomalous nature of the pseudo range and full phase measurements (presence of jumps, rapid changes, etc.) and the divergence of the pseudo range smoothing procedure taking into account the increments of the full phase.
This ensures integrity, taking into account the possible failures of navigation satellites, unacceptable reduction in the signal-to-noise ratio, temporary shadowing of navigation satellites, the effect of anomalous interference and other factors. In addition, the correct change of ephemerical information and the use of identical data in the formation of corrections by each basic receiver are ensured.
2. The integrity monitoring by the values of B-values calculated from the data of all basic receivers that were not rejected at the first stage and operating under the influence of different multi-path. Thus, the integrity is ensured, taking into account the possible impact of intense multi-path and the nature of changes in the differential corrections generated by each basic receiver. Here, the value of the correction to the pseudo range transferred to users, the value of its increment, the presence of jumps and other anomalous changes of these values are monitored (in the communication of the type 1).
3. The output (integral) integrity monitoring on the value of the current coordinate determination errors in the differential solution with the use of corrections and data transmitted to users in the basic receiver.
In civil aviation much attention is paid to the possibility of obtaining benefits from the use of high-precision modern GLONASS/GPS navigation satellite systems at all stages of flight, including precision landing approach. However, the weak power of the satellite signal (by the criterion of weight and power consumption of the transmitter on the satellite) makes the signal dependent on external factors (disturbance of the ionosphere, state of the troposphere, solar flares, magnetic storms, industrial radio interference). Therefore, the important task at the use of GNSS signals is to ensure the integrity of satellite navigation signals.

The measure of integrity is the probability of detecting a failure over the period of time equal to or less than the set one. ICAO standards set high integrity requirements (confidence-building measures) for GNSS signals and their quality (SQM). The autonomous satellite navigation field of GLONASS/GPS does not meet the ICAO integrity requirements at all stages of the aircraft flight. Table 3.2 shows the ICAO requirements for signal characteristics in space.

The traditional approach to assessing the risk of loss of integrity of a GBAS station at precision landing approach of the CATI is as follows. For each subset i of the set of possible failures, it is verified that the GBAS ground subsystem monitors ensure that the following conditions are met [16]

$$P_{IR,i} = P_{PL,i} P_{MD,i} P_{PR,i} \leq P_i. \tag{3.4}$$

Table 3.2. ICAO requirements (RTCA) for signal characteristics in space.

Typical operation	Accuracy (95%)		Integrity		
	Horizontally	Vertically	Time to signal	Signal threshold	Probability of an erroneous signal
Oceanic and remote regions	12.4 nm	NS	2 min	12.4 nm	10^{-7}/h
Intracontinental routes	2.0 nm	NS	1 min	2.0 nm	10^{-7}/h
Airfield area	0.4 nm	NS	30 s	1.0 nm	10^{-7}/h
Non-precision approach	220 m	NS	10 s	0.3 nm	10^{-7}/h
Precision approach CAT I	16 m	7.7 m	6 s	Hor. – 40 m Vert. – 10–15 m	2×10^{-7}/approach
Precision approach of the CAT II	6.9 m	2.0 m	2 s	Hor. – 17.3 m Vert. – 5.3 m	2×10^{-9}/approach
Precision approach of the CAT III	6.1 m	2.0 m	2 s (target is 1 s)	Hor. – 15.5 m Vert. – 5.3 m	2×10^{-9}/approach

Here $P_{IR,i}$ is the risk of loss of integrity for the approach for the subset i of the set of possible failures; P_i is the maximum permissible value of the risk of loss of integrity for the subset i of the set of possible failures; $P_{PR,i}$ is the a priori probability of occurrence of this type of failures per approach; $P_{MD,i}$ is the conditional probability (provided that the failure occurred) that the monitors did not detect the failure and did not exclude anomalous measurements during the maximum permissible alarm time after the anomaly became potentially dangerous; $P_{PL,i}$ is the conditional probability (provided that the failure occurred and was not detected) that the resulting error in determination of the user's position exceeded the level of protection calculated by him.

The main disadvantage of the integrity risk assessment methodology (3.4) is that the final result strongly depends on the value of the a priori probability $P_{PR,i}$ of occurrence of a specific type of failure. The estimates of a priori probability values $P_{PR,i}$ are based on single events and therefore are not reliable. For example, the distortion of the NS navigation signals was observed (at least, it is considered so) only once in 1993 and has not been reliably explained. At the same time, according to (3.4), the smaller is the value of a priori probability $P_{PR,i}$, the "worse" the monitors operate, which parry the influence of these failures on the integrity of the system [17].

3.2.1 The Algorithms to Ensure the Noise Immunity of the Complex Navigation System in the Area of an Airfield by Protection Levels

The necessary characteristics of the integrity of the signals are provided only with the joint use of space groups and functional augmentations to the GNSS. To improve the integrity and continuity in the GBAS ground equipment, it is proposed to use protection levels taking into account the estimate σ_{pr_gnd} based on local noise or blocking interference as a control and in on-board NUE.

The levels of protection in the vertical plane and horizontal plane (VPL and LPL) of the signal in space (SIS) are the upper limits of the confidence interval of the coordinate determination error in relation to the reference point of the ground subsystem: the risk of loss of integrity of the level of protection during the approach should be less than $0.5 * 10^{-7}$ for an approach.

The calculation of protection levels is based on the projection of the estimate of the probability density function of the error in the corrected pseudo ranges to the coordinate area. The risk of loss of integrity for the level of protection is the risk caused by undetected errors in determination of the location in relation to the reference point of the ground subsystem, exceeding the alarm thresholds under the following two conditions [18]:

(a) In normal measurement conditions;
(b) In the measurement conditions with errors.

For no-fault conditions, the vertical protection level (VPL_{HO}) and the horizontal protection level (LPL_{HO}), which involve measurements in the absence of failures, use 2–4 GNSS basic receivers and all satellites and are calculated as follows:

$$VPL_{HO} = K_{ffmd}\sqrt{\sum_{i=1}^{N} s\_vert_i^2 \times \sigma_i^2} \tag{3.5}$$

$$LPL_{HO} = K_{ffmd}\sqrt{\sum_{i=1}^{N} s\_lat_i^2 \times \sigma_i^2}, \tag{3.6}$$

where K_{ffmd} is the product obtained from the probability of error-free detection of a missing;

s_vert_i is the projection of the vertical component for the *i*th satellite (range measuring source);

s_lat_i is the projection of the side component for the *i*th satellite;

N is the number of satellites used in the navigation solution; and i is the satellite index.

The hidden failure includes any unreliable measurement that is not detected by GBAS immediately, due to which a coordinate error is entered into the on-board NUE subsystem.

The vertical protection level (VPL_{H1}) and the horizontal protection level (LPL_{H1}), in the variant where no detected failure is present in one of the GNSS basic receivers, are defined as follows [19, 20]:

$$VPL_{H1} = \max\left[VPL_j\right] \tag{3.7}$$

$$LPL_{H1} = \max\left[LPL_j\right], \tag{3.8}$$

where VPL_j and LPL_j (j = from 1 to MAX $\{M_i\}$) for all GNSS basic receivers of ground subsystems are:

$$VPL_j = |B\_vert|_j + K_{md}\sigma_{vert,H1}, \tag{3.9}$$

$$LPL_j = \left|B\_lat_j\right| + K_{md}\sigma_{lat,H1}, \text{ and} \tag{3.10}$$

$$B\_vert_j = \sum_{i=1}^{N} \left(s\_vert_i \times B_{i,j}\right) \tag{3.11}$$

$$B\_lat_j = \sum_{i=1}^{N} \left(s\_lat_i \times B_{i,j}\right), \tag{3.12}$$

$B_{i,j}$ are differences of corrections without the participation of the *j*th receiver on the *i*th satellite;

K_{md} is the coefficient calculated from the probability of missing of a malfunction of a ground station;

$$\sigma^2_{vert,H1} = \sum_{i=1}^{N} \left(s\_vert_i^2 \times \sigma\_H1_i^2\right), \tag{3.13}$$

$$\sigma^2_{lat,H1} = \sum_{i=1}^{N} \left(s\_lat_i^2 \times \sigma\_H1_i^2\right), \tag{3.14}$$

$$\sigma\_H1_i^2 = \frac{M_i \times \sigma^2_{pr\_gnd,i}}{M_i - 1} + \sigma^2_{pr\_air,i} + \sigma^2_{tropo,i}, \tag{3.15}$$

M_i is the number of reliable $B_{i,j}$ connected with the pseudo range corrections for the *i*th range measuring source.

$$\sigma_{pr\_gnd} = \sigma *_{pr\_gnd} + \Delta\sigma_{pr\_gnd},$$

where $\Delta\sigma_{pr\_gnd}$ is the RMSD of the differential correction error caused by noise.

The RMSD of the ground station contribution to the error of the corrected pseudo range for GPS and GLONASS satellites is calculated by the formula [18, 21]:

$$RMS_{pr\_gnd} \leq \sqrt{\frac{\left(a_0 + a_1 e^{-\theta_n/\theta_0}\right)^2}{M} + (a_2)^2}, \quad (3.16)$$

where

M is the number of installed ground SNS of basic receivers;

n is the *n*th signal source (satellite);

θ_n is the elevation angle for the *n*th satellite.

Tables 3.3 and 3.4 show the accuracy requirements for various systems.

The MSRD of the ground station contribution to the corrected pseudo-range error for SBAS satellites is described by the expression:

$$RMS_{pr\_gnd} \leq \frac{1,8}{\sqrt{M}}(\mathrm{m}). \quad (3.17)$$

The error introduced by the GBAS ground station in the user's measured GNSS and corrected pseudo-range (PR) values is characterized by the standard deviation (SD) of the differential correction (DC) formation error to the PR. σ_{pr_gnd}. The estimates $\sigma^*_{pr\_gnd}$ of this parameter are calculated by the methods of mathematical

Table 3.3 The requirements to GBAS-GPS accuracy

GBAS precision letter symbol	θ_n (°)	a_0 (m)	a_1 (m)	θ_0 (°)	a_2 (m)
A	≥ 5	0.5	1.65	14.3	0.08
B	≥ 5	0.16	1.07	15.5	0.08
C	35	0.15	0.84	15.5	0.04
	5–35	0.24	0	–	0.04

Table 3.4 The requirements to GBAS-GLONASS accuracy

GBAS precision letter symbol	θ_n (°)	a_0 (m)	a_1 (m)	θ_0 (°)	a_2 (m)
A	≥ 5	1.58	5.18	14.3	0.078
B	≥ 5	0.3	2.12	15.5	0.078
C	35	0.3	1.68	15.5	0.042
	5–35	0.48	0	–	0.042

statistics based on the results of processing of test statistics that have information about errors in the formation of the DC [22].

LMCS-A-2000 (GBAS), as well as LMCS of the CATII and CATIII, implemented the algorithms for generation of the estimates $\sigma^*_{pr\_gnd}$ in real time during the operation of LMCS. These estimates are transmitted to consumers in the communication 1.

The typical implementations of DC to pseudo range for NS no. 2, 3, 5, 10, 15, 24 and the estimates $\sigma^*_{pr\_gnd}$ corresponding to these DC are shown in Figs. 3.16a, b, 3.17a, b, 3.18a, b.

The qualitative analysis of the presented implementations shows that, in most cases, the estimates $\sigma^*_{pr\_gnd}$ are maximum at the NS rise and set. With an increase in the angle of elevation $\overline{El}$, as a rule, the error in the formation of the DC decreases, as evidenced by a decrease in the parameter σ_{pr_gnd} at large angles of elevation.

Since the error in the formation of the DC and its statistical characteristics are unknown, then the accuracy of the estimate formation $\sigma^*_{pr\_gnd}$ can only be estimated by indirect methods. To do this, according to the results of post-processing of measurement records of the BR from the LMCS, the errors of formation of the DC are isolated. To isolate these errors, other methods and test statistics are used than for parameter estimation σ_{pr_gnd}.

It is known that for a Gaussian random variable the probability of falling beyond the area limited by the values $\pm 3\sigma$ is 0.0027.

For a qualitative assessment of the accuracy of the formation σ_{pr_gnd} the estimates of the errors in the formation of the DC obtained by the post-processing algorithms (green color), and the boundary values $\pm 3\sigma^*_{pr\_gnd}$ (red color) are derived further on the same graph.

The parameter estimates σ_{pr_gnd} are formed in accordance with the algorithms implemented in the LMCS [23].

Such combinations of estimates for NS No. 2, 3, 5, 10, 15, 24 are presented in Figs. 3.19a, b, c, 3.20a, b, c. The qualitative analysis of these graphs shows that the errors in the formation of the DC cross the boundary $\pm 3\sigma^*_{pr\_gnd}$ quite rarely.

For the quantitative assessment of the quality of assessment $\sigma^*_{pr\_gnd}$ formation, it is possible to use an empirical estimate of the probability of falling beyond a set area $\pm 3\sigma_{pr\_gnd}$

$$P^*\left\{\left|\varepsilon_{pr}\right| > 3\sigma_{pr\_gnd}\right\} \approx \frac{m}{n}. \tag{3.18}$$

Here m is the number of outputs for the specified area; n is the total number of counts.

For NS No. 2, 3 and 5, the probability (3.18) is 0.0033, 0.0034 and 0.0032, respectively. The obtained empirical values of probabilities differ from the theoretical values by not more than 20%.

Thus, the estimate $\sigma^*_{pr\_gnd}$ in the LMCS, taking into account the limited amount of data used to calculate (3.18), is formed with a sufficiently high accuracy. These estimates can be used to determine the accuracy class of LMCS.

Check of the accuracy class of LMCS

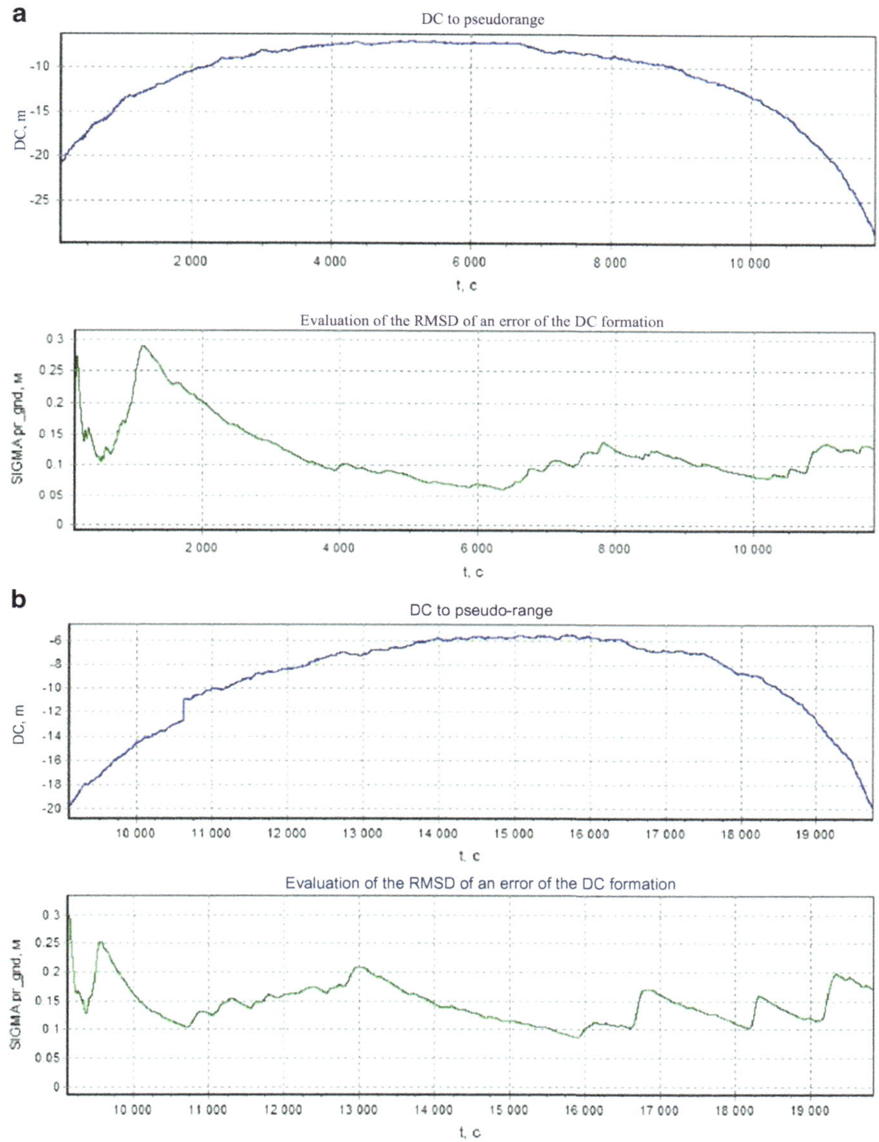

Fig. 3.16 **a** DC to pseudo range for NS No. 2 and RMSD error estimate **b** DC to pseudo range for NS No. 3 and RMSD error estimate

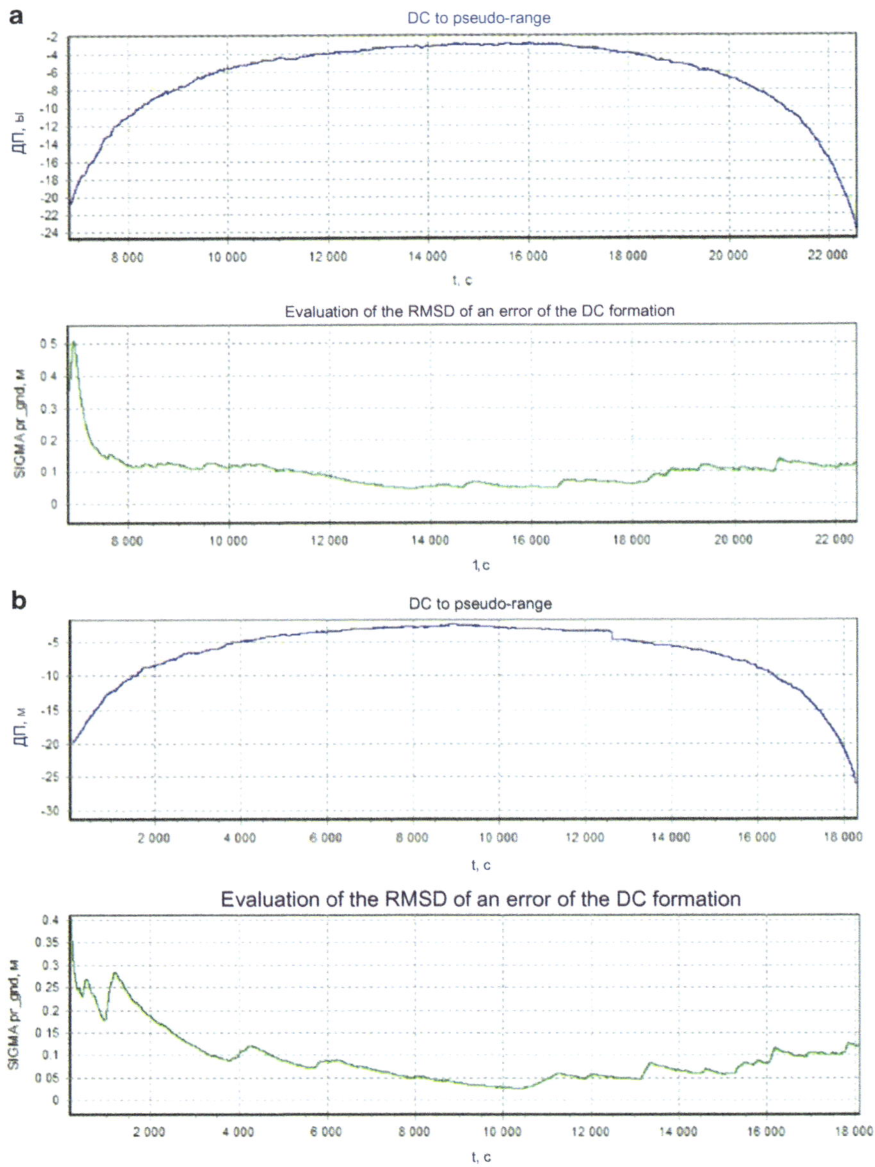

Fig. 3.17 **a** DC to pseudo range for NS No. 5 and RMSD error estimate **b** DC to pseudo range for NS No. 10 and RMSD error estimate

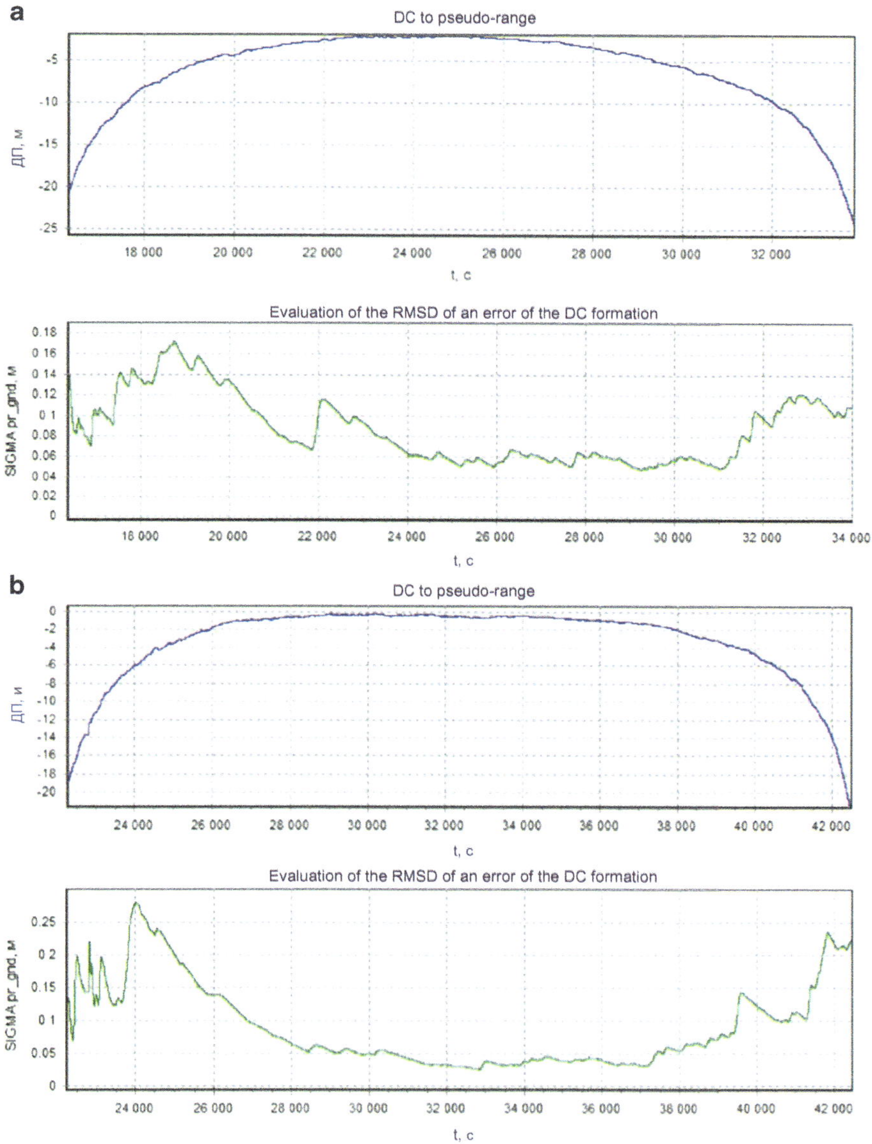

Fig. 3.18 **a** DC to pseudo range for NS No. 15 and RMSD error estimate **b** DC to pseudo range for NS No. 24 and RMSD error estimate

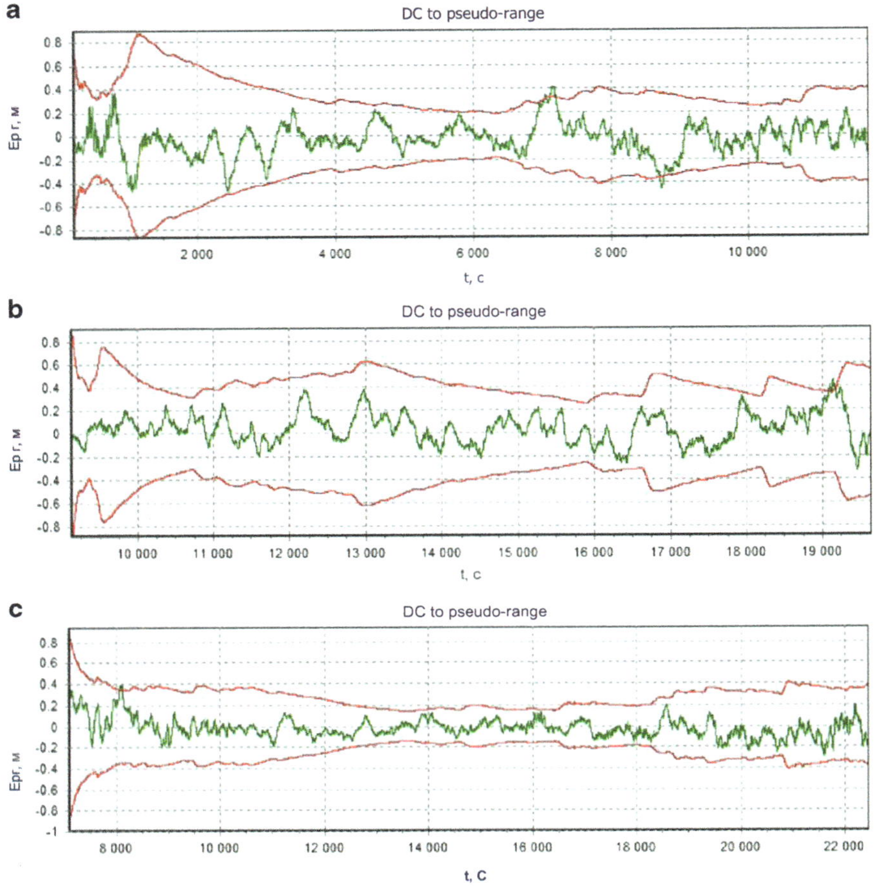

Fig. 3.19 **a** The error of DC formation for NS No. 2 **b** The error of DC formation for NS No. 3 **c** The error of DC formation for NS No. 5

According to the SARPs of the RMSD, the error of the corrected user PR introduced by the ground subsystem GBAS and determining its accuracy class for the n.th GPS and GLONASS NS are described by the expression [12, 19]

$$\sigma_{pr\_gnd,n} = \sqrt{\frac{\left(a_0 + a_1 e^{\frac{El_1}{El_0}}\right)^2}{M} + a_2^2}, \tag{3.19}$$

where M is the number of LMCS basic receivers; El_n– is the angle of elevation of the nth NS; a_0, a_1, a_2 and El_0 are the parameters given in Tables 3.3 and 3.4 for various accuracy classes of the ground subsystem (GAD) [24].

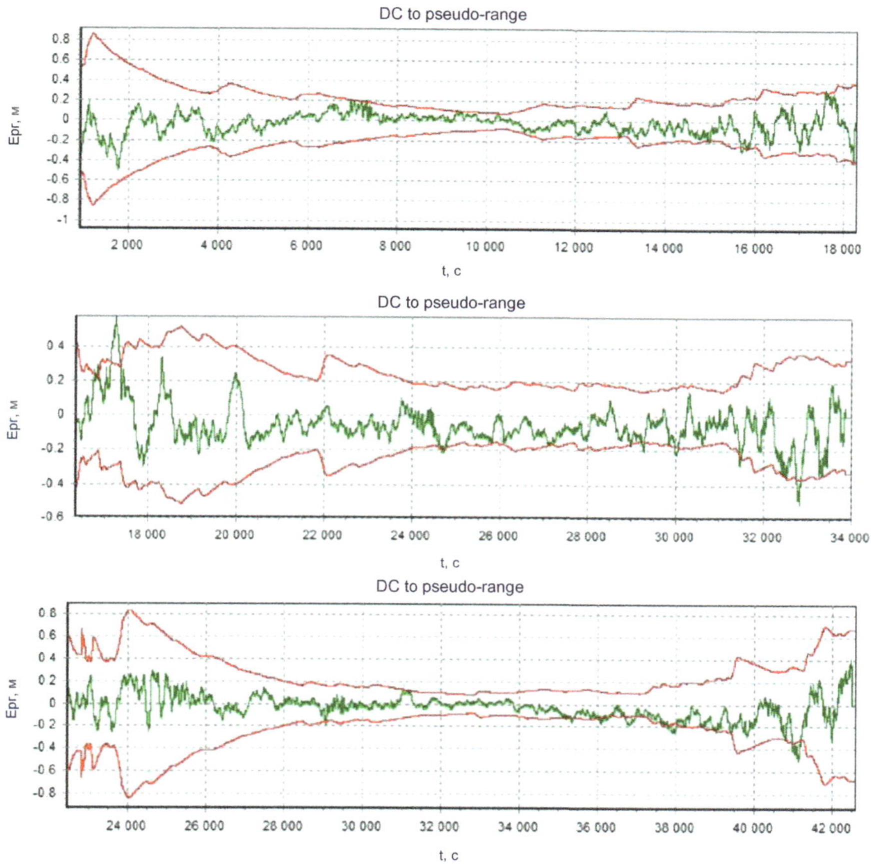

Fig. 3.20 **a** The error of DC formation for NS No. 10 **b** The error of DC formation for NS No. 15 **c** The error of DC formation for NS No. 24

Note 1. The requirement to the accuracy of GBAS stations is determined by the letter indication of the accuracy of the equipment and the number of GNSS control receivers.
Note 2. Functional extensions to the GNSS (GBAS) and satellites do not take into account the failures of the aircraft onboard equipment.

If the actual RMSD of the error in the formation of the DC transmitted to the user does not exceed the calculated values σ_{pr_gnd} (3.19), obtained for a given accuracy class, then the LMCS corresponds to this accuracy class.

The decision is made on the basis of the aggregation of the DC of all the observed NS, not rejected by the algorithms of quality control of the DC formation for a given time interval. However, only the estimates of this parameter $\sigma^*_{pr\_gnd}$ are available for observation, which are formed with an error.

Investigations of the accuracy characteristics of estimation algorithms σ_{pr_gnd} have shown that for the estimate $\sigma^*_{pr\_gnd,k}$, formed in the kth instant of time, in the first approximation it is possible to write

$$\sigma^*_{pr\_gnd,k} = \sigma_{pr\_gnd,k} + \varepsilon_k, \tag{3.20}$$

where ε_k is the error of parameter estimation σ_{pr_gnd}, the statistical dynamics of which is described by the model of the form

$$\varepsilon_{k+1} = f_\varepsilon \varepsilon_{k+1} + \gamma_\varepsilon n_{\varepsilon,k}, \tag{3.21}$$

$$f_\varepsilon = e^{-\alpha T}, \quad \gamma_\varepsilon = \sigma_\varepsilon \sqrt{1 - f_\varepsilon^2},$$

α, σ_ε are the parameters characterizing the correlation time and the RMSD of the process (3.21), $n_{\varepsilon,k}$ is a sample of a Gaussian random variable with zero AA and a single RMSD. Here, for simplicity of the formula dependencies, the index of the NS number is omitted.

When sampling estimates (3.20) in a stationary area of a sufficiently large volume, the number of estimates that exceed the true value of the parameter σ_{pr_gnd} is approximately equal to the number of estimates that lie below the true value of this parameter. Thus, if the number of estimates $\sigma^*_{pr\_gnd,k}$, exceeding the calculated threshold of the accuracy class (3.19), is less than the number of assessments lying below this threshold, then this accuracy class is confirmed; otherwise the parameter σ_{pr_gnd} does not correspond to the declared accuracy class.

Confirmation of the LMCS accuracy class is performed in the following order: for a set test interval the current estimates $\sigma^*_{pr\_gnd,n,k}$ of all the observed NS not rejected by the control algorithms are projected onto the plane of the elevation angle—the parameter σ_{pr_gnd}. According to the test results, a cloud of estimates $\sigma^*_{pr\_gnd,n,k}$ which characterizes the accuracy of LMCS is formed. In the same graph, the dependencies (3.19) are derived, which are the boundaries of the corresponding accuracy class.

In Figs. 3.21 and 3.22 in red the dependence (3.19) corresponds to the class A, green to the class B, and blue to the class C. The assessment cloud $\sigma^*_{pr\_gnd,n,k}$ is formed from the results of processing of daily data of LMCSA-2000.

It follows from the analysis of the presented data that the tested LMCS corresponds to the accuracy class GAD B2, both for the GPS system and the GLONASS system: all samples of estimates $\sigma^*_{pr\_gnd,n,k}$ fall well below the threshold of the class A, and there are insignificant falling of these estimates beyond the threshold values of the class B. Formally the quantity of estimates $\sigma^*_{pr\_gnd,n,k}$, exceeding the class C thresholds, is less than 50% of the total sample size; however, at elevation angles from 5° to 15°, this condition is not satisfied.

If the estimate of the accuracy class is more rigid (to test the most unfavorable conditions and make the decision about the entire LMCS by these tests), then this LMCS does not correspond to the GAD C class [25] (Figs. 3.23, 3.24).

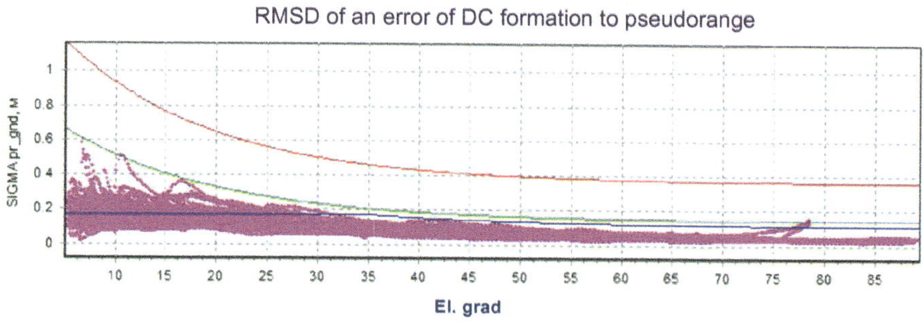

Fig. 3.21 RMSD of an error of DC formation, GPS

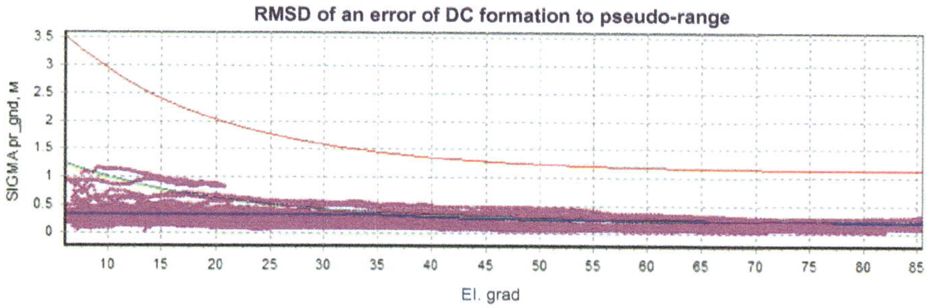

Fig. 3.22 RMSD of an error of DC formation, GLONASS

Conclusion: The obtained calculated values of the accuracy estimate of the differential corrections lie below the level of requirements specified in the ICAO standards for satellite landing systems.

To ensure the integrity control of the VDB data link, it is proposed to use a VDB monitoring receiver (board simulator).

Conclusion: B-values are below the requirements specified in the ICAO standards for satellite landing systems.

An important factor in ensuring the integrity of the satellite landing system is ensuring the required power of the VDB transmission in the landing zone (20 nm). Figure 3.25 provides the estimate of the level of intensity of the radio field of VDB of LMCS-A-2000 when carrying out flight tests in the Gromov Flight Research Institute and in a/p Sheremetyevo (Fig. 3.26).

Conclusion: The intensity of the radio field during flight tests lies in the range from −35 to −99 dB/W/m^2. The integrity VDB data link is provided.

Conclusion: The accuracy characteristics of the satellite landing system comply with the ICAO requirements and are confirmed in flight tests in the Gromov Flight Research Institute.

Improving the integrity of the satellite navigation landing system is aimed at elimination of the possibility of the use of unreliable data by an on-board computer.

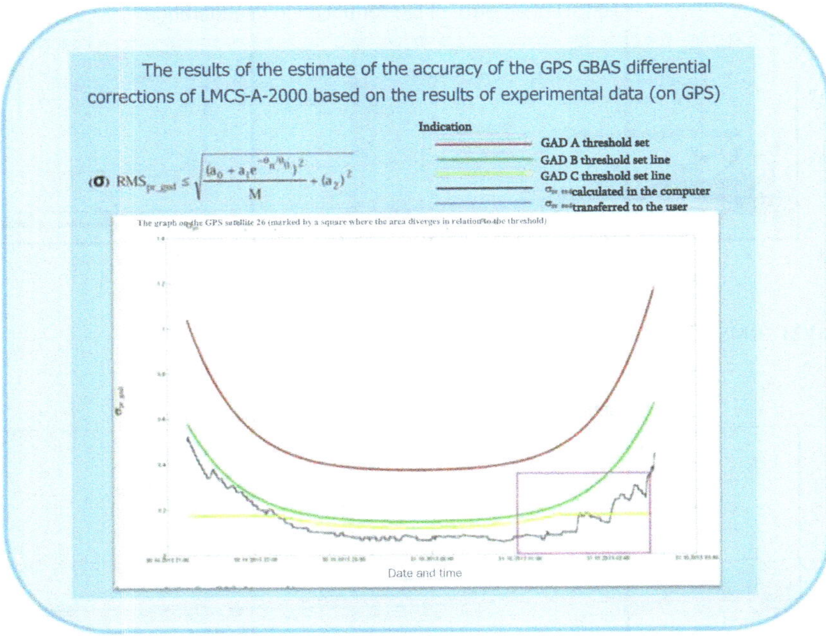

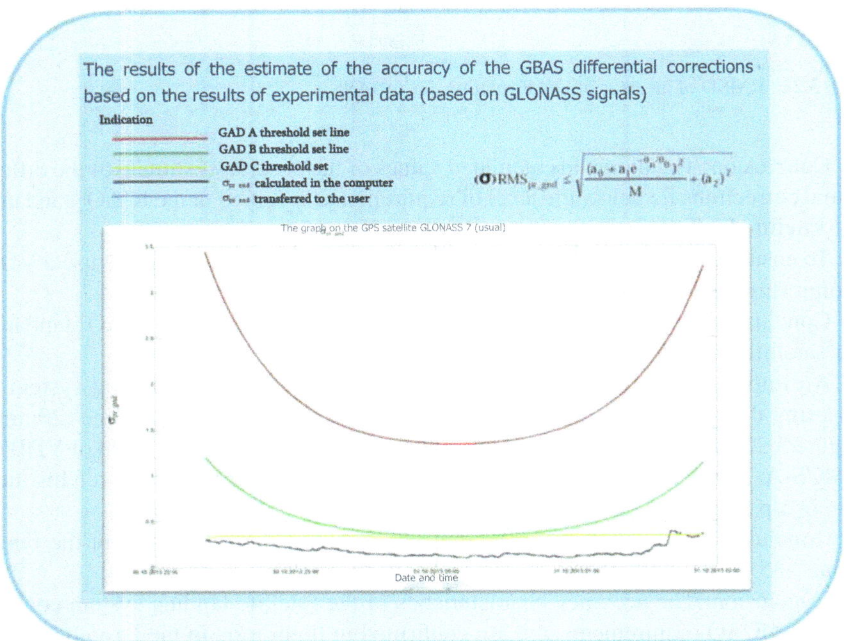

Fig. 3.23 The results of the estimate of the accuracy of the GBAS differential corrections based on the results of experimental data (based on GPS and GLONASS signals)

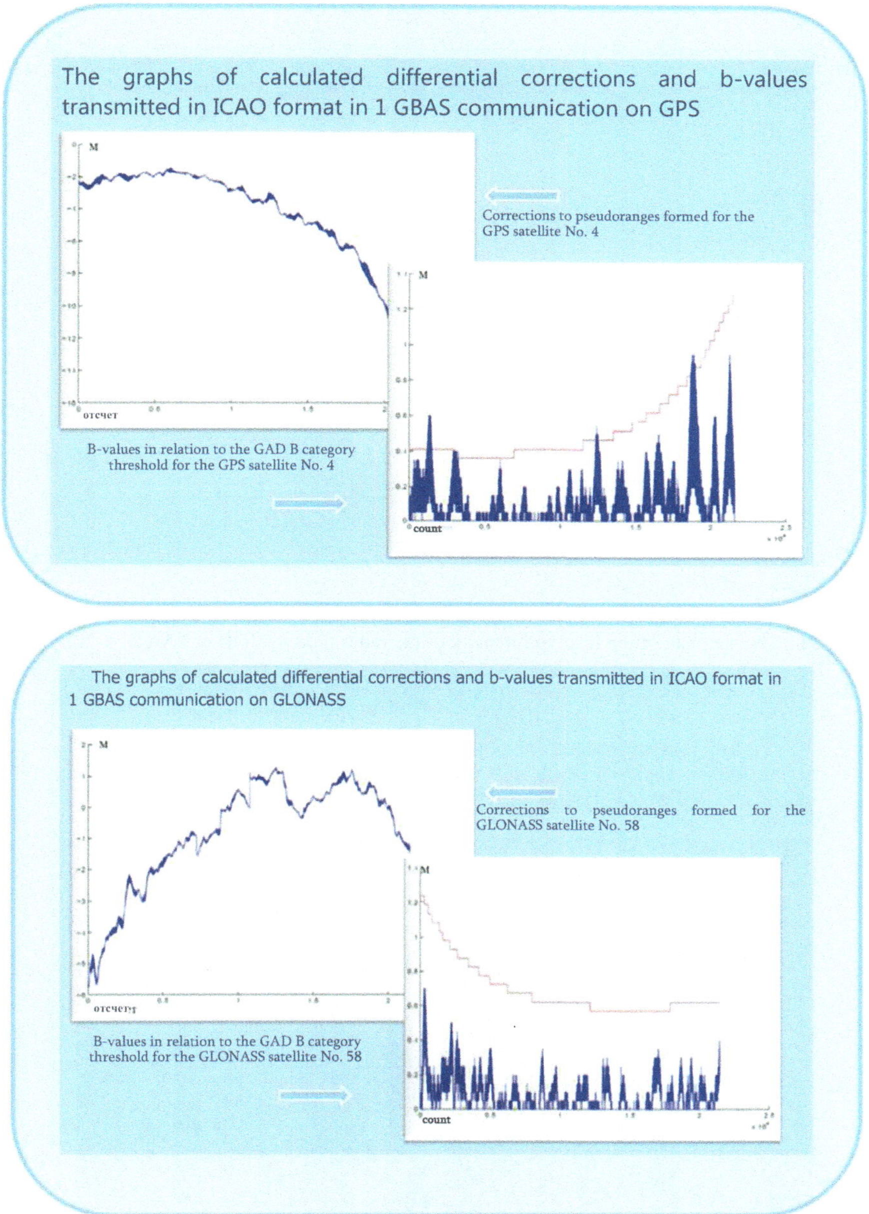

Fig. 3.24 The graphs of calculated differential corrections and B-values transmitted in ICAO format 1 GBAS communication in GPS and GLONASS

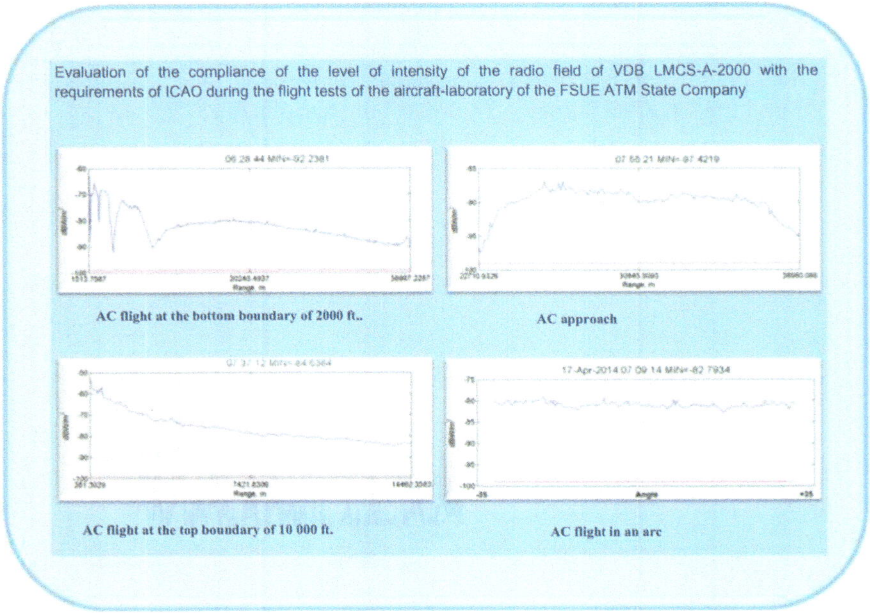

Fig. 3.25 The estimate of the level of intensity of the radio field of VDB of LMCS-A-2000

Flight assessment of the accuracy characteristics and continuity of the DGNSS at the landing approach of the Il-976 aircraft with the use of the GBAS 11/111 station and GLS on-board equipment

Mode No	PEE, 95%,(M)		FTE, 95%, [M]		TSE, 95%, [M]		Mode time [c]	Piloting mode
	horizontal	vertical	horizontal	vertical	horizontal	vertical		
Mode 1	0,8	0,5	6.0	4,3	6.0	4,3	100	automatic
mode 2	0,7	1,4	5,4	4,8	5,4	5,0	102	automatic
mode 3	1,4	0,3	4,2	4,5	4,4	4,5	108	automatic
mode 4	0,9	0,9	5,2	3,2	5,3	3,3	96	automatic
mode 5	2,2	1,9	4,0	2,2	4,6	2,9	111	automatic
mode 6	1,8	1,8	4,3	3,5	4,7	4,5	113	automatic
mode 7	1,5	0,8	4,8	3,9	5,0	4,0	105	director
mode 8	1,4	1,1	4,1	4,2	4,7	4,4	100	director
mode 9	0,8	1,0	2,4	4,1	2,5	4,2	110	automatic
mode 10	1,3	0,9	5,3	2,1	5,5	2,3	112	automatic
mode 11	0,8	1,3	4,3	3,2	4,4	3,5	70	director
mode 12	1,2	1,5	5,2	2,8	5,3	3,2	75	director
mode 13	0,7	2,0	6,0	1,9	6,0	2,8	72	director
mode 14	1,4	0,4	5,2	4,8	5,4	4,8	66	director
mode 15	0,5	0,6	5,9	3,0	5,9	3,1	78	director
mode 16	1,4	1,3	4,5	3,1	4,5	3,4	70	director
mode 17	1,8	1,9	3,0	4,6	3,5	5,0	75	director
mode 18	0,4	0,9	4,4	2,2	4,4	2,4	72	director
mode 19	2,0	1,9	4,5	4,2	4,9	4,7	85	director
mode 20	0,5	1,2	5,5	3,9	5,5	4,5	63	director

Fig. 3.26 The flight evaluation of the accuracy characteristics and continuity of the DGNSS

LMCS stations are proposed to implement a set of measures to reduce the possibility of undetected failure. Such measures are [20, 26]:

- The use of differential navigation mode to exclude common errors of the control station and the NUE (e.g., errors in the clock);
- The installation of several GNSS separated antennas to protect against pseudo range determination errors caused by uncorrelated multiple reflections (multipath). For these purposes the LMCS BOM values are calculated:
- The use of receivers with different gates of the correlator to assess the quality of the satellite signal (SQM) and protect against distortion of the PRS at the satellite;
- The analysis and evaluation of received navigation data (DQM);
- The analysis and evaluation of navigation measurements (MQM) with the use of code and phase measurements, and simultaneous measurements at two-frequency receivers to estimate the ionospheric delay;
- The use of ground-based and on-board navigation data with only identical ephemeris checksums. This feature is applied for the first time in the aviation differential mode and excludes the use of LMCS and NUE of a different set of satellites;
- The use of geodetic connection of a reference point of LMCS to assess the accuracy of the differential mode and the formation of corrections to the pseudo range to range-measuring sources;
- The use of remote monitoring receiver to assess the integrity of the radio channel;
- The use of the model of the ionosphere and troposphere to estimate the signal delay in the propagation medium;
- The estimation of the scaling factor for the difference between the ephemeris of the LMCS location and the location of an aircraft;
- The use of the estimate (one sigma) of the accuracy of the differential corrections for the calculation of the protection levels in the on-board NUE;
- The estimation of the integrity of the satellite navigation signal using 100 and 30 s smoothing intervals of PR;
- The use of hot standby hardware and software of LMCS.

The obtained accuracy and radio-technical characteristics of LMCS are subject to assessment using the high-precision metrological measuring instrument GFTC, which has a certificate of a measuring device type of LMCS (GBAS) of the Federal Agency for Metrology.

The ground and flight test complex (GFTC) is designed to determine own geodetic coordinates and high-precision coordinates of moving objects and stationary equipment placement points, coordinates of local monitoring and correction stations (LMCS) and quantify the parameters of GNSS signals (GLONASS/GPS / Galileo), and space (SBAS) and ground (GBAS) functional extensions.

GFTC is the software and hardware complex that provides the function of a measuring instrument when carrying out ground and flight tests of functional extensions to the Global Navigation Satellite System (GNSS) during commissioning and operation, and is designed to quantify the parameters generated and transmitted by functional extensions to GNSS during tests, as well as during their operation.

The list of the main estimated parameters [24] is as follows:

- The accuracy of determination of the location in the GNSS differential mode by NUE equipment,
- The intensity of the radio field created by the functional extensions to GNSS,
- The compliance with the accepted (measured) from the functional extensions and parameters specified by the air navigation services provider.

Additional estimated parameter is the estimate of the deviation from the line of a given path (reference path).

The differential mode for satellite GPS and GLONASS groupings signals implies the accuracy of units of meters for navigation applications. The basic GNSS receiver is located at a geodetically linked point in the WGS-84 or PZ-90.11 system. The correction can be determined from the difference in geodetic coordinates of the location and coordinates calculated by satellite. Differential corrections are used by local users (350 km) to adjust the coordinates of a user in the differential mode. The differential technology is correct if the errors are systematic and caused by external to NUE causes.

The main systematic sources of errors are:

- Selective access errors (for GPS only) are man-made errors introduced into satellites for secrecy purposes. The errors of this type in the pseudo range at the level 1σ are about 30 m. Currently, selective GPS access in the C/A code in GPS satellites has been removed. There are no selective access errors in GLONASS satellites.
- Ionospheric delays are group signal propagation delays, whose mean value is 20–30 m during the day and 3–6 m at night. When operating at two frequencies, this effect is greatly reduced by applying the inverse quadratic dependence of the delay on frequency.
- Tropospheric delays are signal propagation delays caused by the lower layers of the atmosphere. As for satellites with small elevation angles, these delays reach 30 m, they are quite stable and simulated. The variations in the refractive index can cause the difference (between the reference station and the user) in signal delays of up to 1–3 m for low satellites.
- An ephemerical error is the difference between the actual location of the satellite and its location, predicted by orbital data. Usually this difference is quite small, less than 3 m.
- Satellite clock error is the difference between the satellite clock time and the time predicted by orbital data. The on-board time generators of the GPS and GLONASS satellites are not synchronized with each other. Ground GPS and GLONASS control stations set time corrections for their satellites with respect to ground UTC time, accepted for each of the systems that are entered in the messages transmitted by the satellites. The user reads the data from the satellites taking into account the entered time corrections and synchronizes them with his own on-board time.

The differential mode fully compensates satellite clock errors, while the GNSS receiver of the signal user and the ground station use the navigation data from the same satellites. Ephemerical errors (30 m and more) are compensated by the differential mode. Selective access errors affecting the GPS time adjustment are also compensated by the differential mode, except for those corrections that lose their validity after a certain time.

For users located near the station, the signal propagation geometry from satellites is close enough and the compensation is almost complete.

When the user moves away from the reference station, the satellite signal paths through the ionosphere and the troposphere will differ significantly, since atmospheric discontinuities can cause delays of various sizes.

Due to the fact that the distance between the user and the base station is different, it causes an error in the GLONASS / GPS differential mode (spatial decorrelation). The decorrelation increases with the increase in distance between NUE and GBAS (LMCS). The differential mode specifies the coordinates and increases the integrity to 400 km.

The GNSS differential mode provides the function of monitoring of the integrity and quality of a satellite navigation signal and transmits this information to users. The coordinates of the point are obtained by calculating them from distances to the NSC.

The spherical surfaces of the position correspond to the measured distances (Fig. 3.27). The coordinates of the point of intersection are determined from the three position surfaces, where the receiver is located. At its core, this is a geometric inverse spatial spacing [27].

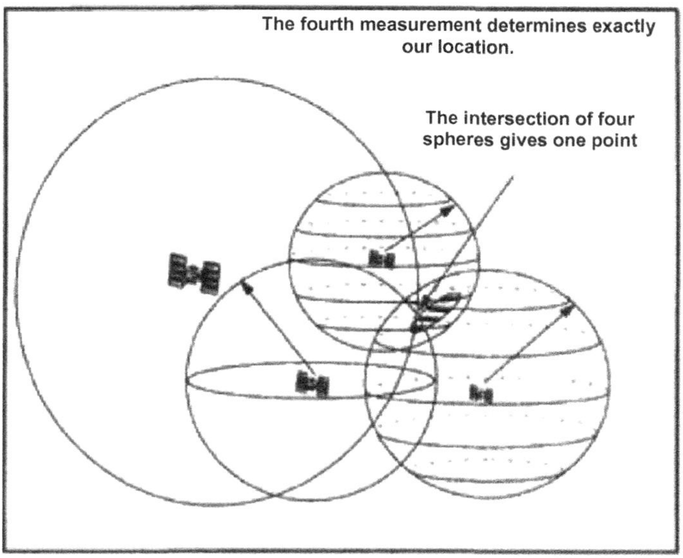

Fig. 3.27 The principle of determination of the location by the satellite method

To obtain the required characteristics in the area of an airfield and at precision landing approach, it is proposed to use the functional extension of LMCS.

LMCS-A-2000 (GBAS) in combination with on-board equipment provides zone navigation (P-RNAV) and precision landing approach RNP 0.02/40 in accordance with international standards: the accuracy of less than 4.0 m; integrity is $1–2 \times 10^{-7}$; continuity is $1–8 \times 10^{-6}$.

Based on the research we can draw the following conclusions [27, 28]:

1. The creation of multi-frequency multi-system (GLONASS/GPS/Galileo) ground-based MF/MC CAT II/III GBAS (multi-frequency/multi-constellation category II/III GBAS) functional systems allows to solve fundamentally the GBAS integrity problem in the presence of ionospheric anomalies (ionospheric storms) due to the transmission of two-frequency estimates of ionospheric delays of the LMCS BR to the users.
2. Multi-frequency systems allow to improve the quality of the solution of the navigation task in the DM due to the use of smoothing of the measured pseudo ranges taking into account the increments of the linear combination of complete phases, free from the influence of the ionosphere. So at strong multi-path, errors in solving the navigation task in the DM at using the usual smoothing procedure can reach 6–8 m. When using a two-frequency smoothing procedure in a similar situation, the errors in solving the navigation task in the DM do not exceed 0.8 m.
3. The use of Galileo and Beidou along with the already used GLONASS and GPS GNSS signals significantly improves the availability and continuity of GBAS systems in areas of airfields, in which part of the sky is shaded by closely located mountain ranges.
4. Duplication of information on the state of GNSS accuracy at workplaces of the landing controller and flight manager from LMCS also improves the integrity of navigation, since if the VHF data line fails, LMCS in accordance with the Federal Aviation Regulations of ATM will immediately inform the crew of an aircraft by voice communication Dispatcher-Pilot.

In addition to the methods improving the integrity of the satellite navigation system through the use of optical UV navigation sensors based on optimal placement of ground equipment and three-dimensional visualization of the terrain on an aircraft board and monitoring of MQM, DQM and SQM described earlier, it is also proposed to estimate B-values, σ_{pr_gnd} and estimates of local interference to the SRNS signal.

3.2.2 Improving the Noise Immunity of the Complex Navigation System in the Area of an Airfield and at Precision Landing Approach by Detecting External Influences on the GNSS Signal

Due to the fact that the airport is the object of highly concentrated sources of radio emission and is located most often in the area of large megacities, the presence of noise interference on the GNSS signals is very probable. In existing solutions of differential systems this circumstance is not taken into account. And taking into account the transience of aircraft landing operations, this may lead to the transfer to board of unreliable estimates of the accuracy of the differential corrections and, as a result, to the overestimated levels of protection (especially vertical VPL) in the on-board NUE algorithms.

To eliminate the possibility of an undetected error in the formation of differential corrections to PR, the following is proposed.

In order to assess the impact of local (noise) interference, the algorithm for estimating and accounting of the coefficient σ_{pr_gnd} was developed based on the analysis of experimental data from Domodedovo Airport.

When radio systems work in the LMCS zone, the quality of the formation of differential corrections (DC) to pseudo range (PR) decreases due to an increase in the PR measurement error. At the same time, there is a loss of tracking of the NS signals, especially at low elevation angles.

To control the power of the received signal, the communication [EC] is used, which contains the C/A signal-to-noise ratio (SNR) for all the NS defined in the current [SI] communication. The existing SNR algorithms are reduced to checking of exceeding in the current time point $t_k = kT_e$ of the set threshold and making the decision

$$\lambda^*_{SNR,k} = \begin{cases} 0, SNR_k > h_{SNR} \\ 1, SNR_k \le h_{SNR} \end{cases}, \qquad (3.22)$$

where usually $h_{SNR} = 20\ldots30$ dBHz. In order not to overweight the formulas, here the index of the NA number is omitted.

This monitor rejects those NS for which the breakdown of DTS tracking (delay tracking scheme) and the auto-tuning function of the radio signal parameters are possible, but does not detect impairment in the measurement accuracy of PR and (or) PP under external influence.

The signal-to-noise ratio varies with the angle of elevation of the NS. In Fig. 3.28 green dots show the current signal-to-noise ratio in the normal mode, depending on the angle of elevation for all observed NS for 4 h of BR operation.

The empirical curve is drawn through these points (a cloud of points) (shown in Fig. 3.28 in red), which delivers the minimum of the sum of squares of deviations from the experimental samples. This curve is set by the equation of the form

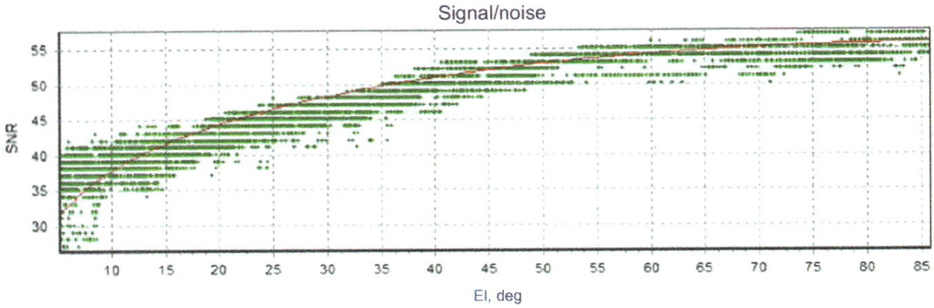

Fig. 3.28 Current signal-to-noise ratio in normal mode

$$SNR(El) = 20\log\left\{\frac{\sin(\pi - 2El)}{\pi - 2El}\right\} - SNR_{90}, \qquad (3.23)$$

where $SNR_{90} = 56$ dB Hz is the signal-to-noise ratio for NS at the zenith; the function $\sin(x)/x$ describes the pattern of the receiving antenna.

It is possible to use the differences between the SNR estimates transmitted by the receiver in the [EC] communications and the calculated SNR as test statistics for detecting external influences, in accordance with (3.23). As an illustration, Figs. 3.29, 3.30 and 3.31 present typical implementations of SNR estimates for NS No. 4, 13, 16 in normal mode. From these graphs it can be seen that in normal mode the difference between the SNR estimate and the calculated value (3.23) does not exceed ± 10 dB.

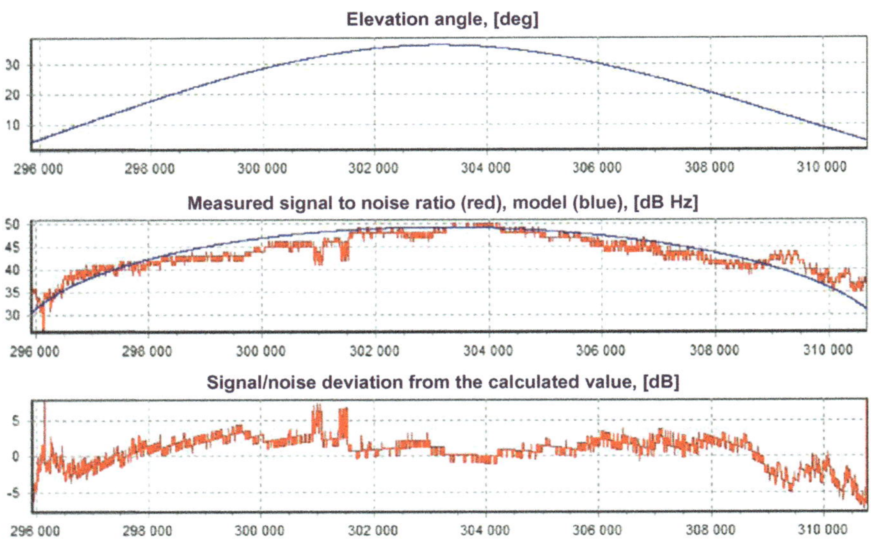

Fig. 3.29 Typical implementations of SNR estimates for NS No. 4 in normal mode

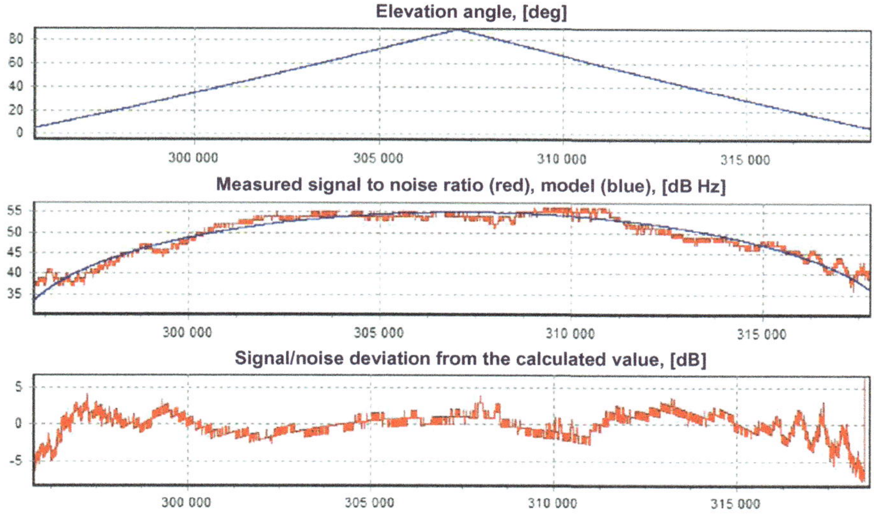

Fig. 3.30 Typical implementations of SNR estimates for NS No. 13 in normal mode

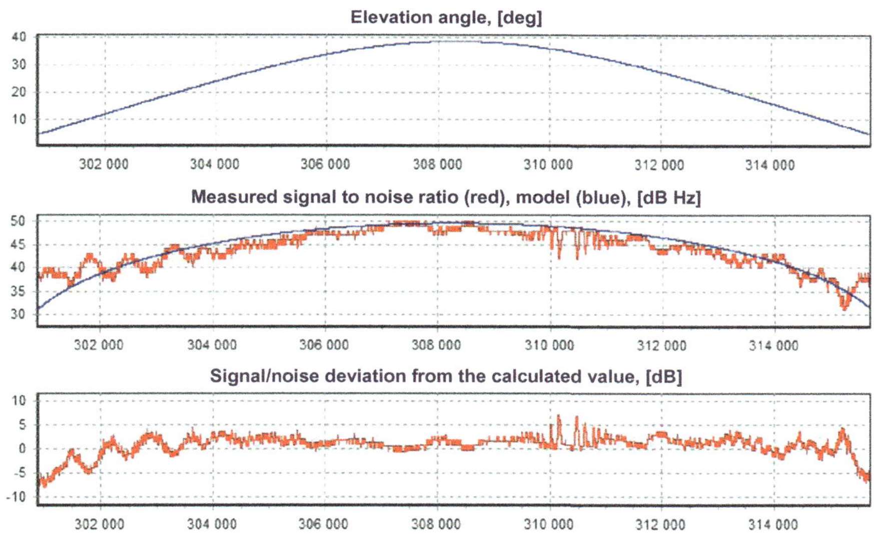

Fig. 3.31 Typical implementations of SNR estimates for NS No. 16 in normal mode

Figures 3.32 and 3.33 show typical implementations of SNR estimates for NS No. 8, 14 under conditions of abnormal external influence. From these graphs it can be seen that after the occurrence of the anomaly ($t > 276{,}000$ s) the difference between the SNR estimate and the calculated value (3.22) increases to 10–15 dB.

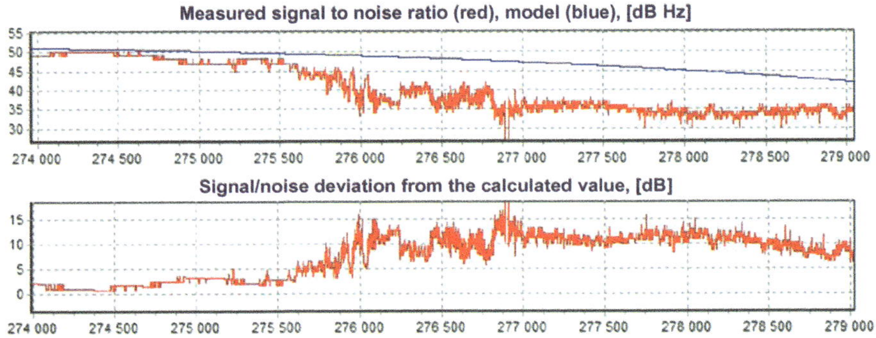

Fig. 3.32 Typical implementations of SNR estimates for NS No. 8 under conditions of abnormal external influence

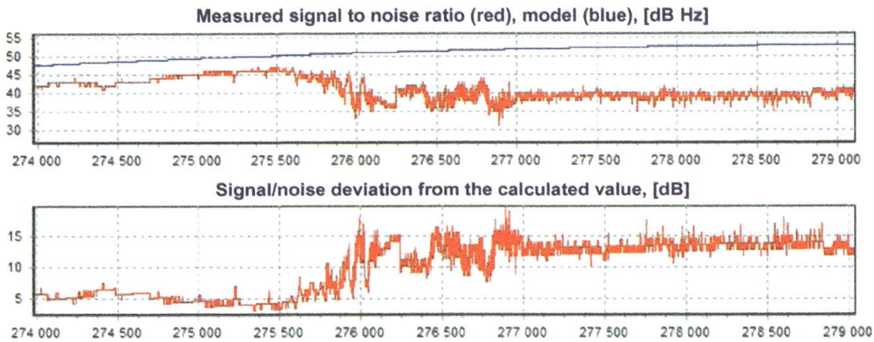

Fig. 3.33 Typical implementations of SNR estimates for NS No. 14 under conditions of abnormal external influence

The decision on the presence of external interference is made by two thresholds $h_1 = h_{SNR}$ and $h_2 = h_{SNR} + \Delta h$, and the decisive rule is [29]

$$\lambda_{P,k}^{*} = \begin{cases} 0, \left(\delta SNR_k < h_2 \wedge \lambda_{SNR,k-1}^{*} = 0\right) \vee \left(\delta SNR_k < h_1 \wedge \lambda_{SNR,k-1}^{*} = 1\right) \\ 1, \left(\delta SNR_k \geq h_2 \wedge \lambda_{SNR,k-1}^{*} = 0\right) \vee \left(\delta SNR_k \geq h_1 \wedge \lambda_{SNR,k-1}^{*} = 1\right), \end{cases} \tag{3.24}$$

where $\delta SNR_k = SNR_k - SNR_k^{*}$, SNR_k is the calculated signal-to-noise ratio (3.23); SNR_k^{*} is the evaluation of the signal-to-noise ratio transmitted by the receiver.

This decisive rule provides that when the value δSNR_k decreases below the threshold value $h1$, the threshold value is replaced by $h_2 > h_1$. Usually $\Delta \text{h} = 3.5$ dB. In the process of programming and debugging of the algorithm for detection of external interference, it was found that the best threshold values are $h_1 = 5$ and $h_2 = 10$ dBHz.

To study the statistical characteristics of errors in PR measuring, two-frequency combinations of PP (PPC) are used of the form

$$\frac{\left(\mu_n^{(2/1)}+1\right)\varphi_n^{(1)}(t_k)-2\varphi_n^{(2)}(t_k)}{\mu_n^{(2/1)}-1}, \tag{3.25}$$

in which the delay in the ionosphere has the same sign as in code measurements. Here $\varphi_n^{(1)}$, $\varphi_n^{(2)}$ are the measurements of PP in the ranges L_1, L_2, respectively, $\mu_n^{(2/1)}$ is the frequency ratio.

The use of these combinations in PR smoothing algorithms allows to increase the smoothing time from 100 to 1000 sec. The error in the formation of such smoothed PR is substantially less than the error of smoothed PR with the time constant of 100 s, i.e., their difference, in the first approximation, can be mistaken for the error in the formation of smoothed PR. Accordingly, the difference between the initial PR and the smoothed PR with the smoothing time of 1000 s can be taken as the measurement error of the PR. Realizations of errors in the formation of PR for NS 8, 10 and 11 are shown in Figs. 3.34a, b, c, respectively.

As the results of research have shown, the external influence leads to an increase, mainly in the noise component of the measurement errors of the PR. So for NS No. 10 after the occurrence of the anomaly, the error rate increased from 0.7 … 1 m to 2 … 3 m. At the same time, in smoothed PR, the effect of external interference is not so much observed, the error of smoothed PR increases from 0.2 m to 0.3 m and a significant increase in these errors is observed at low angles of elevation, which is typical for the normal mode [29].

Thus, the influence of external interference mainly leads to an increase in the noise component of measurement errors. The use of smoothing algorithms with the time constant of 100 s reduces the level of the fluctuation component of noise by about 10 times. On average, errors in the formation of DC due to external influence increase by 10–20%.

The external interference is detected by the difference in SNR estimates transmitted by the receiver in communications [EC], and the calculated SNR is in accordance with decision rule (3.24). For NS in which these anomalies are detected, DC is formed in a regular manner, but at the same time the parameter σ_{pr_gnd} increases by 10–20%.

If the numerical values of the formula (3.23) given here, the thresholds of rejection and increase in σ_{pr_gnd} are approximate and should be clarified during the operation of LMCS.

The special place in ensuring the integrity of the differential mode is the control of the ionospheric gradient, especially during the period of solar activity.

Figure 3.35 shows how to estimate the iono gradient during the ionospheric "storm", the formation of NOTAM and the RAIM forecast.

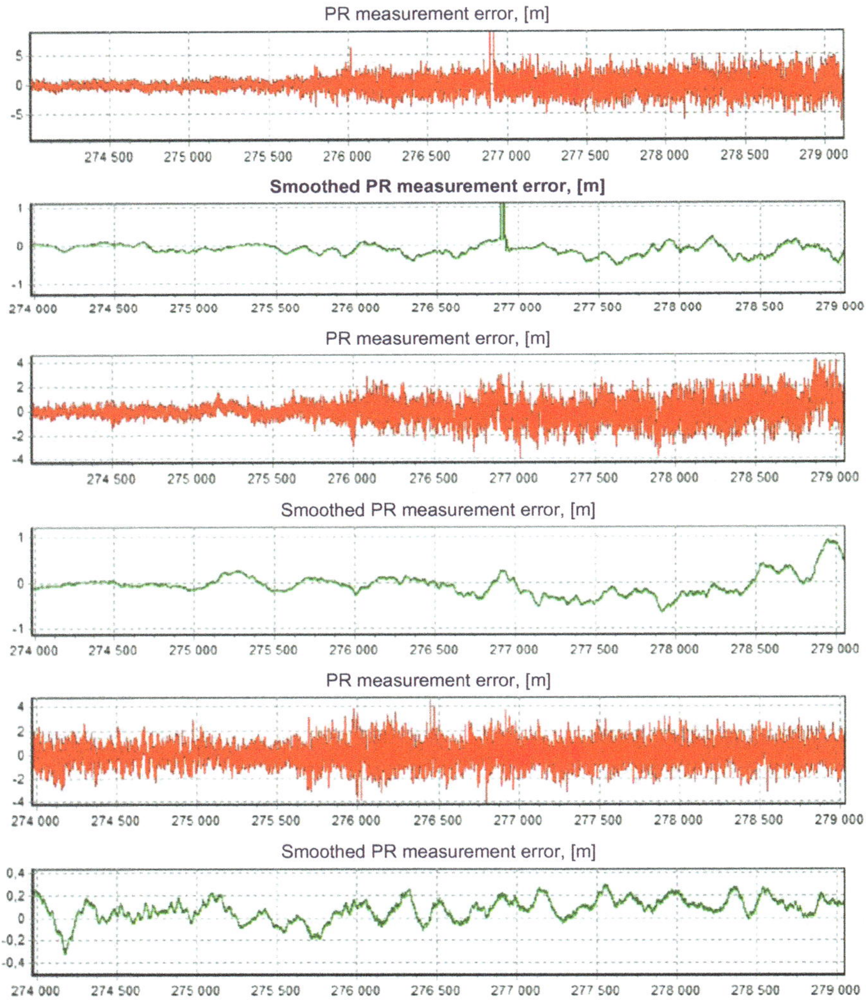

Fig. 3.34 **a** Realization of errors of DC formation for NS No. 8 **b** Realization of errors of DC formation for NS No. 10 **c** Realization of errors of DC formation for NS No. 11

3.2.3 Detection of Anomalous State of the Ionosphere (Ionospheric Storm)

When analyzing the influence of the anomalous state of the ionosphere (ionospheric storm) on the quality characteristics of the solution of the navigation task (NT) in differential mode (DM), the most unfavorable scenario is considered when the storm is spread from an AC side. After the front overtakes the piercing point of the ionosphere by the NS-AC line of sight, the differential error of this NS will

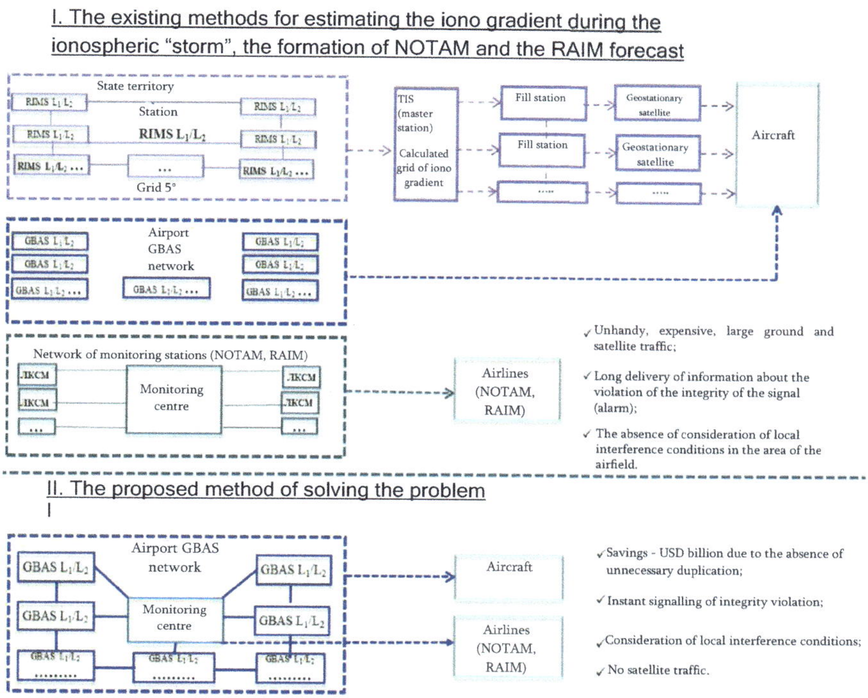

Fig. 3.35 The methods for estimating the iono gradient during the ionospheric "storm", the formation of NOTAM and the RAIM forecast

increase until the front of the ionospheric storm reaches the corresponding LMCS (GBAS) piercing point. At this interval, LMCS cannot detect abnormal delays in the ionosphere. The corrected pseudo range (PR) error will be determined by the distance between an aircraft and the LMCS and the parameters of the ionospheric storm. The model of the ionospheric storm is shown in Fig. 3.36. This model is known as CONUS (Contiguous United States). For the case when the width of the ionospheric storm w is greater than the distance between an aircraft and the LMCS, the maximum error of the corrected pseudo range due to the ionospheric storm is determined by the expression [30]

$$\varepsilon_{pr,z} = m_g g D, \tag{3.26}$$

where g is the spatial gradient of the ionospheric storm [mm/km]; D is the distance between an aircraft and LMCS [m]; $m_g = 10^{-6}$ is the scale factor.

For elevation angles other than 90°, the corrected pseudo range error can be recalculated using the appropriate display functions; for example,

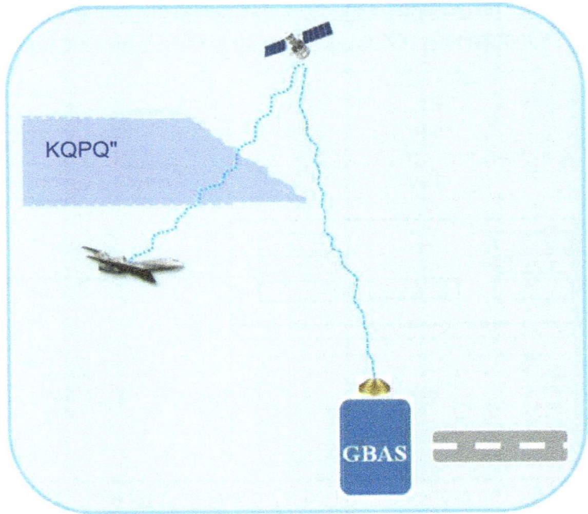

Fig. 3.36 Ionospheric storm

$$\varepsilon_{pr}(El) = \varepsilon_{pr,z}\frac{\varepsilon_{pr,z}}{\sqrt{1-\left(\frac{R\cos(El)}{R+h}\right)^2}}, \tag{3.27}$$

where $R = 6370$ кМ is the radius of the earth; $h = 420$ кМ is the average height of the ionosphere.

With single-frequency measurements at one point, the ionospheric storm can be detected only from the rate of change of the ionospheric delay. The rate of change of the ionospheric delay V_{ion}, the spatial gradient of the ionospheric delay g and the rate of propagation of the ionospheric storm V for zenith NS are connected by the correlation

$$V_{ion,z} = m_g g V. \tag{3.28}$$

In Figs. 3.37 and 3.38, the graphs of ionospheric storm detection for the minimum detectable rate of the ionospheric storm $V_{ion,\min}$, equal to 0.02 and 0.01 m/s, respectively, are presented. Above the curve there is a detection area of the ionospheric storm, and below is the area of non-detection. As it follows from the analysis of the data of the drawings, the most critical from the point of view of detection is a slowly propagating storm with a rate of less than 100 m/s. In this case, at $V_{ion,\min}$, equal to 0.02 m/s, an ionospheric storm with the spatial gradient of more than 200 mm/km is detected. To detect an ionospheric storm with the spatial gradient of less than 200 mm/km, the requirements for the detection of the anomalous rate of lag change in the ionosphere are increased ($V_{\text{ion,min}}$about0.01 m/s).

$$\mathrm{V}_{\text{ion,min}} = 0.02\ \mathrm{m/s}$$

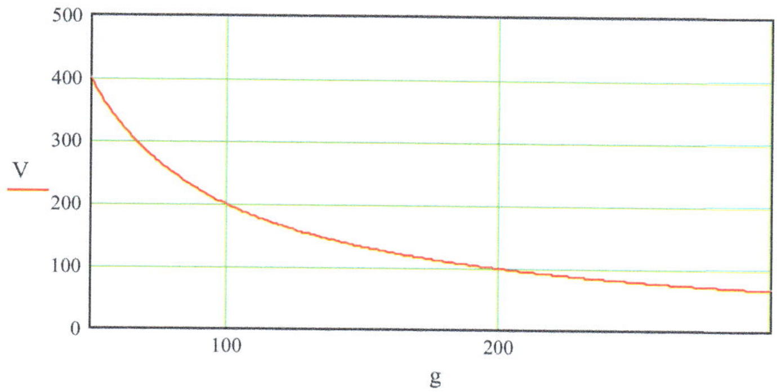

Fig. 3.37 The graph of ionospheric storm detection for 0.02 m/s

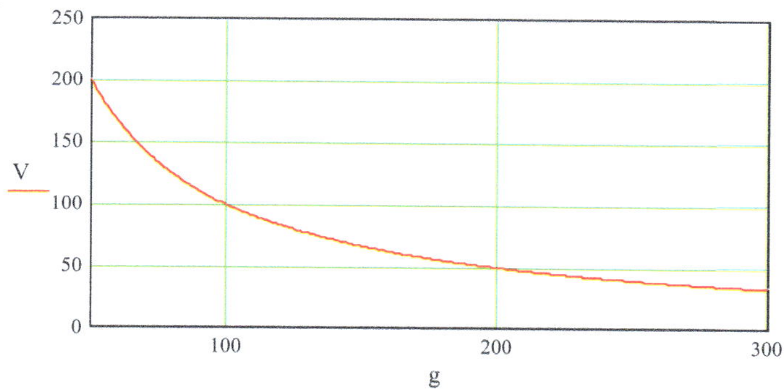

Fig. 3.38 The graph of ionospheric storm detection for 0.01 m/s

$$V_{\text{ion,min}} = 0.01\ \text{m/s}$$

At the set $V_{ion,\min}$, the probability of a false alarm P_{ffmd} and the probability of an undetected failure P_{md} , we can formulate the requirements to the RMSD σ of the test statistics of the detection algorithm for the anomalous rate of change of delay in the ionosphere [31]

$$V_{ion,\min} = \left(k_{ffmd} + k_{md}\right)\sigma, \tag{3.29}$$

where k_{ffmd} is the multiplier determined from the set probability of a false alarm P_{ffmd} and the distribution law of test statistics; k_{md} is determined by a set value of the probability of undetected failure P_{md}.

At $\text{P}_{\text{ffmd}} = 1,9 \times 10^{-4}$ and $P_{md} = 10^{-3}$, the values k_{ffmd} and k_{md} are, respectively, equal to 3.62 and 3.63. At $V_{ion,\min}$, equal to 0.01 m/s, from (3.29) it follows $\sigma = 0,0015\text{M/C}$.

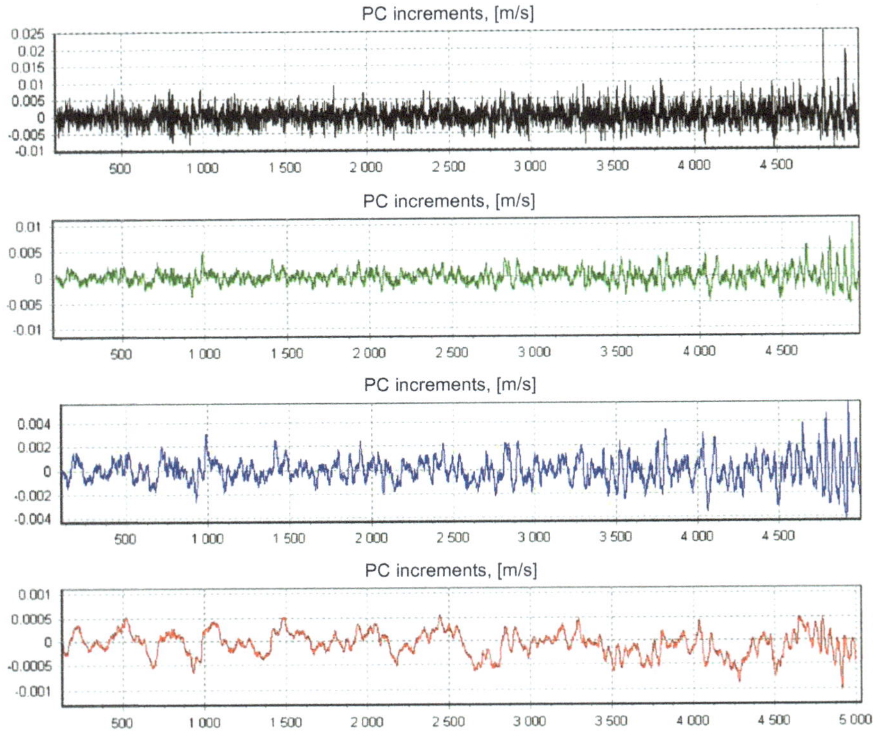

Fig. 3.39 Increment of phase corrections

For NS with low elevation angles with the same parameters of ionospheric storm detection, the requirements for the RMSD of the test statistics are lower (σ about 0.005 m/s).

Figure 3.39 shows typical implementations of test statistics for the algorithms of detection of the anomalous rate of change of the delay in the ionosphere, which use phase correction increments (PC).

The graphs show them in the following order: from top to bottom: the increments of the PC and the smoothed increments of the PC with a filter with the time constant of 5, 10 and 100 s.

Observations at the SP output with the time constant of 100 s have a minimum RMSD. However, they do not satisfy the detection time (about 100 sec). SP with the time constant of 10 sec satisfies the requirements for detection accuracy and time (see Fig. 3.40).

The increments of the PC, without smoothing, time constant are 5, 10, 100 s.

In addition to combinations of phase measurements, combinations of code measurements smoothed by SP with different time constants are used to detect the anomalous ionosphere. So for the SP with the time constants of 100 sec and 30 sec, a DSIGMA monitor is implemented.

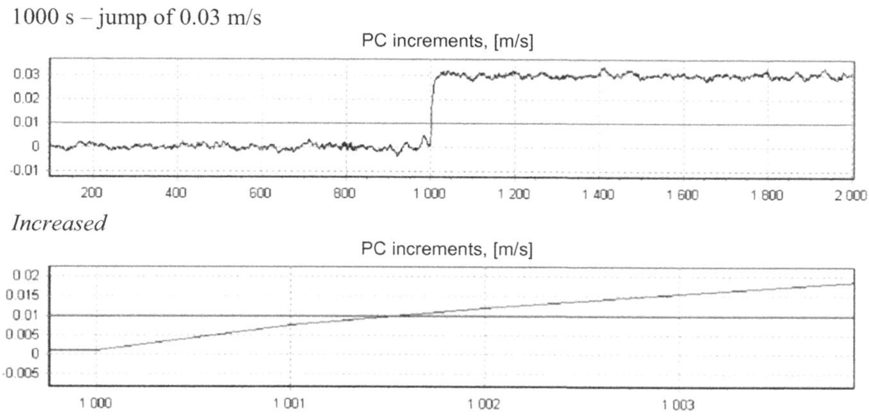

Fig. 3.40 Increment of phase corrections

Figures 3.41, 3.42 and 3.43 show the implementation of test statistics of phase detection algorithms for the anomalous ionosphere (IONO) of the DSIGMA monitor with different parameters of the anomalies. The same graphs show the rates of change in the ionosphere V_{ion} and the delay in the ionosphere due to the anomalous state of the ionosphere D_{ion} .

In the first case (see Fig. 3.41), the limiting case is considered when the parameters of the ionospheric storm model significantly exceed the real one. The second case (see Fig. 3.42) is typical for NS with low angles of elevation, and the third case is for NS close to the zenith.

As it follows from the data analysis in Figs. 3.41, 3.42 and 3.43, the IONO monitor is triggered when the ionospheric delay is $D_{ion} \leq 1$ m . The DSIGMA monitor detects anomalies of the ionosphere in the order of 3–5 m.

However, for IONO-type phase monitors, it is necessary that the receiver is stationary and located at a point with known coordinates.

For a DSIGMA monitor, such restrictions are not required. The only restriction is that an aircraft should not move along the propagation line of the ionospheric storm with the rate of the storm [31].

3.3 An Experimental Confirmation of the Increased Accuracy of Navigation and Landing Using the Local Ground-Based Functional Extension to GBAS (LMCS-A-2000) of the I–III Category of ICAO

The purpose of the tests was the flight test on the functioning of the prototype of GBAS ground station of the Category II/IIIA developed by RDPF Spektr under

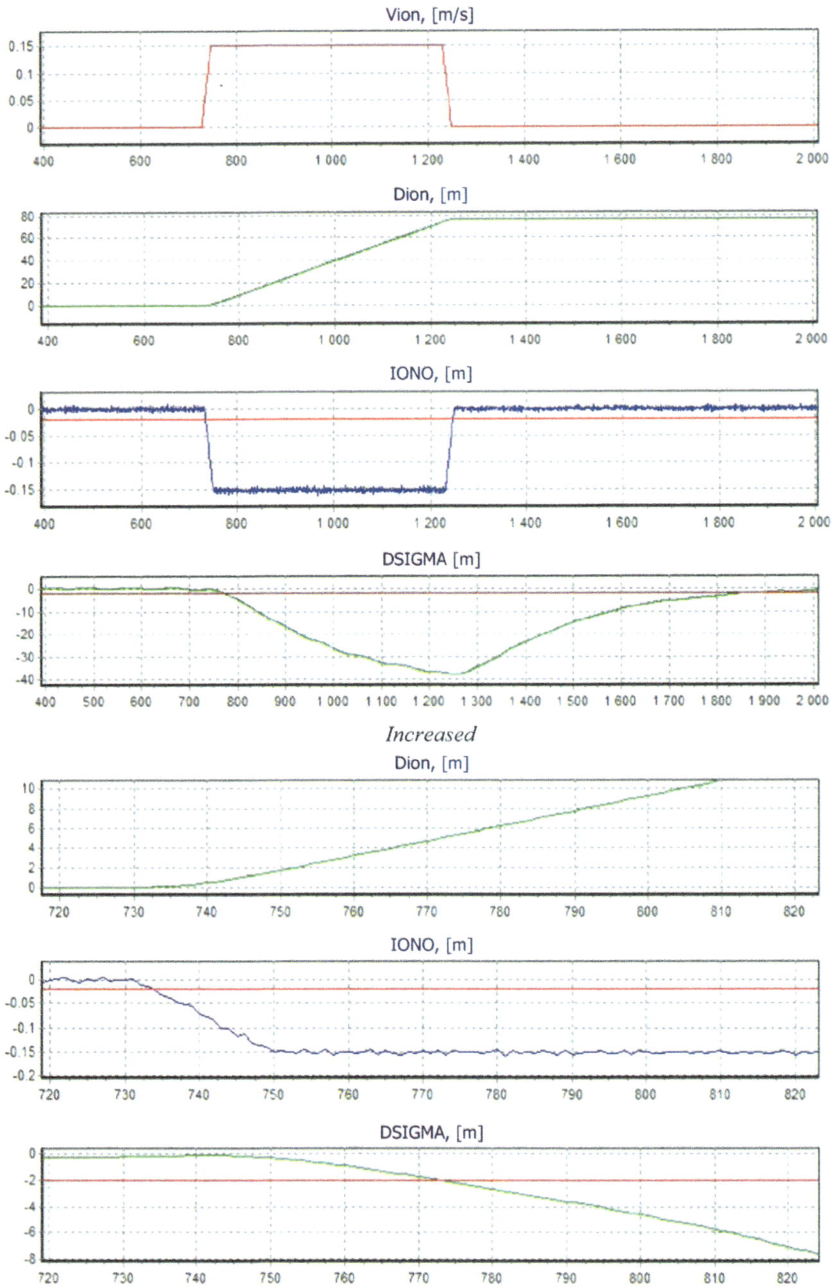

Fig. 3.41 The parameters of the ionospheric storm model exceeding the real one

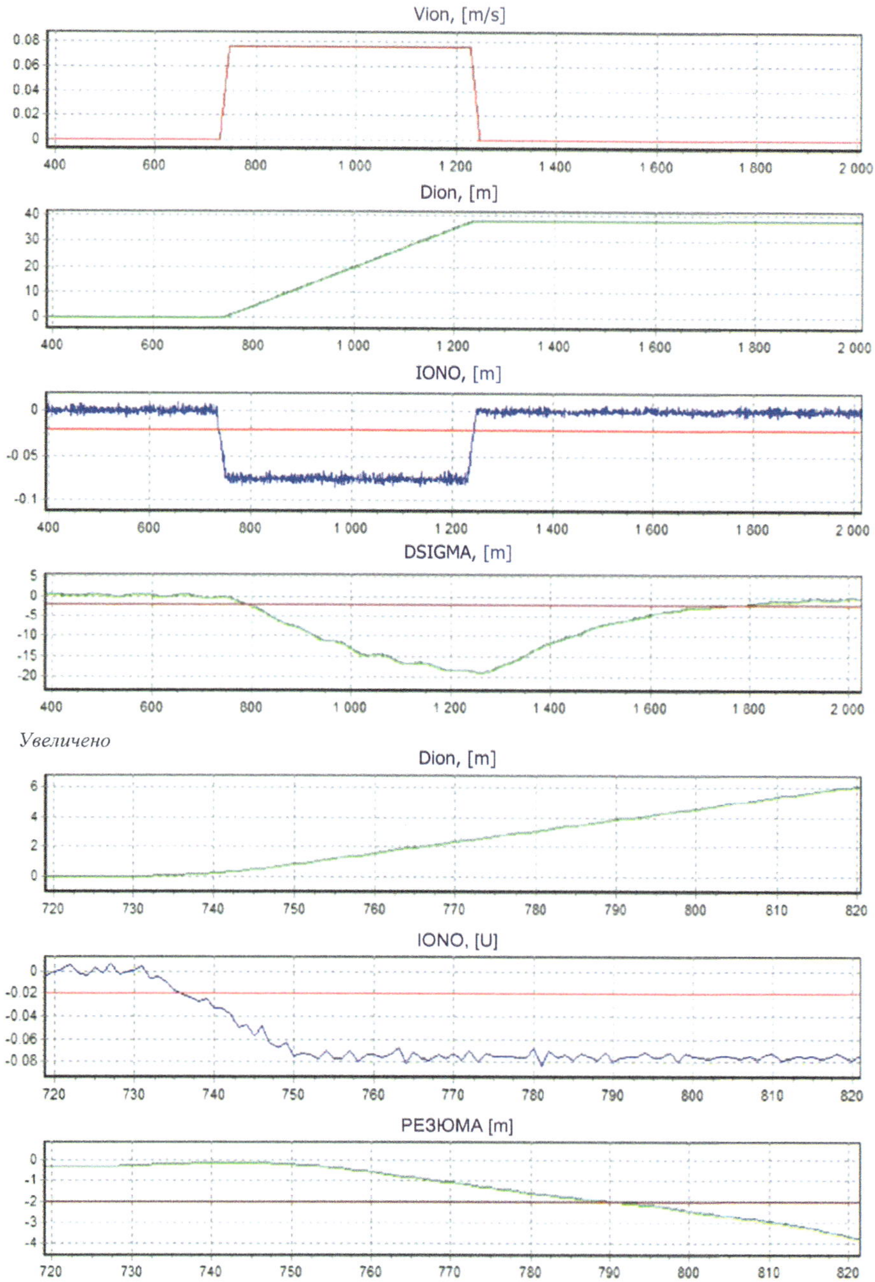

Fig. 3.42 The parameters of the ionospheric storm model, NS with low elevation angles

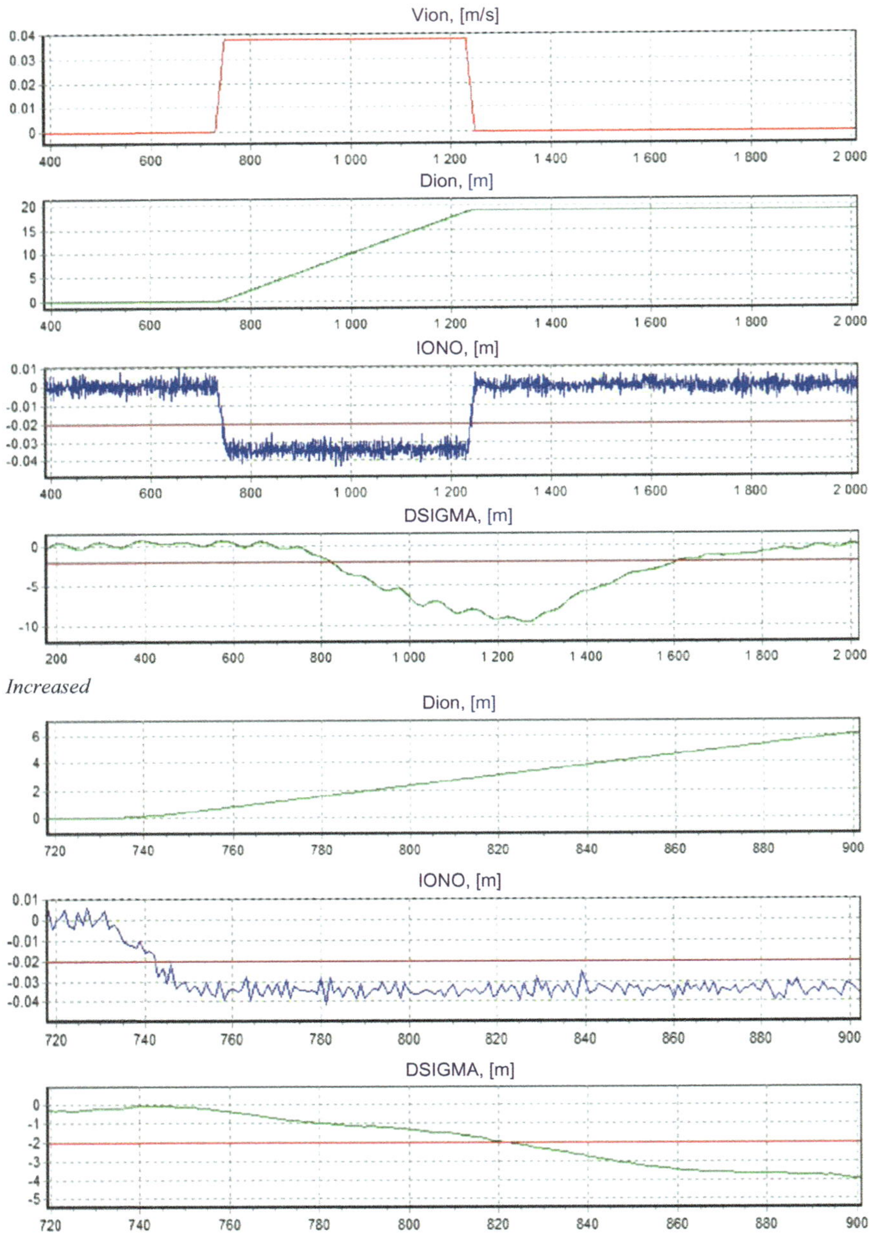

Fig. 3.43 The parameters of the ionospheric storm model, NS close to the zenith

conditions of actual interference of an operating airfield, when flying in an airfield area and precision landing approach of the TU-154 No. 85317.

The GBAS ground station is designed to provide precision landing approach and navigation procedures. In accordance with the ICAO requirements, the GBAS provides positioning and landing services. The feature of the GBAS application is the independence of a set landing trajectory from the influence of external electromagnetic effects and distance from the landing point within a set GBAS coverage area.

The GBAS ground station of the Category IIIA is a functional extension to the ground-based GNSS and is intended to form and transmit in real time to aircraft (via the VDB channel in the range of 108.00–117.95 MHz) and to ground users (via wired, fiber optic, satellite and/or Internet data lines) the following information [32]:

- Differential corrections to pseudo ranges, the rate of change of corrections,
- The information on the status of the signals of the GLONASS and GPS orbital constellations in the station coverage area,
- The information on the admissibility of aircraft performing typical operations using GNSS data in the differential and autonomous mode of operation of the on-board receiver, taking into account the requirements for accuracy, integrity, availability, readiness and continuity of the satellite signal and the time before warning, defined by the RNP for a specific standard operation,
- Optional forecasting of the level of availability of navigation signals of GLONASS and GPS satellite constellations in the station coverage area,
- Quality control of the navigation satellite signal,
- Integrity monitoring parameters (numbers of rejected satellites) when receiving signals from GLONASS and GPS navigation satellites that are in view of the LMCS receivers,
- Additional navigation parameters in SARP's ICAO format on GBAS,
- The station monitoring data.

The hardware and software complex of the Category III GBAS ground station equipment developed by LLC RDPF Spektr (KNTA 466,539.009) is located at the Ramenskoye airfield, consisting of:

(a) KOI building (7th floor, laboratory 93):

  - Reference station GBAS of the category III with antenna-feeder systems GNSS and data link (VDB);
  - The console of an engineer;

(b) CCC-2, tower:

  - The airfield flight manager console with displaying information on allowed/prohibited flight modes on the GNSS signals (GLONASS + GPS or GPS);

- The landing dispatcher console with the display of information on the permission/prohibition of the execution of the final approach stage on the GNSS signals (GLONASS + GPS or GPS).

The diagram of the Ramenskoye airfield with an indication of the locations of the category III GBAS equipment is shown in Fig. 3.44.

The location of SNS antennas and VDB data link on the roof of the building is shown in Fig. 3.45.

The location of the flight manager console and the landing controller console on the CCC-2 tower is shown in Fig. 3.46. Figure 3.47 shows the block diagram of connections of the III category ground GBAS elements.

The specially equipped Tu-154 aircraft No. 85317 was used (Fig. 3.48) as a flying laboratory in flight tests.

Connection diagram of the on-board GLS equipment consisting of:

- Differential data reception equipment—DDRE,
- On-board multipurpose system (landing)—OMS-L.

A solid-state drive of the GBAS type was used as the on-board recorder. The on-board trajectory measurement complex—OTMC combined with the inertial system I-21—was used as a device control of reference trajectories (Figs. 3.49 and 3.50).

The landing signals, formed by GLS, entered the standard automatic on-board system installed on the Tu-154 aircraft—ACSU-154 M, the characteristics of which provide the possibility of landing of the category III.

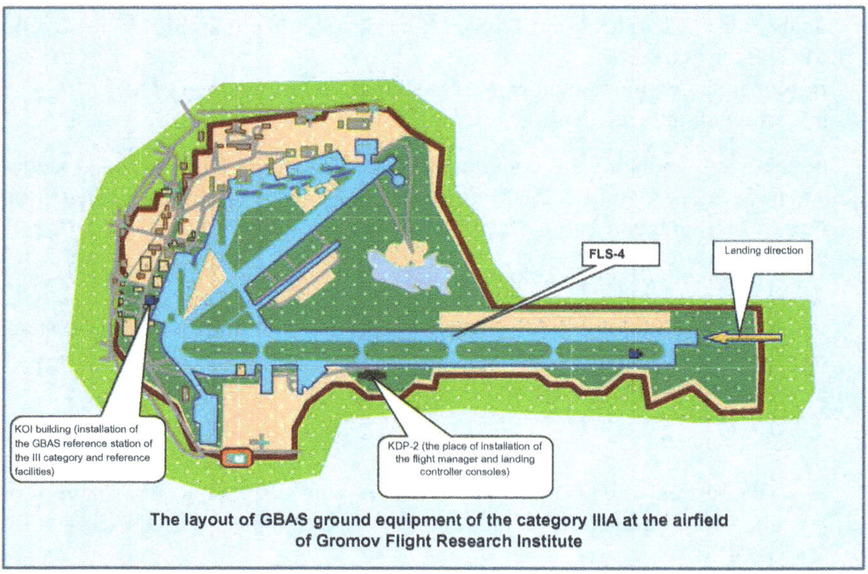

Fig. 3.44 The layout of GBAS ground equipment at the airfield

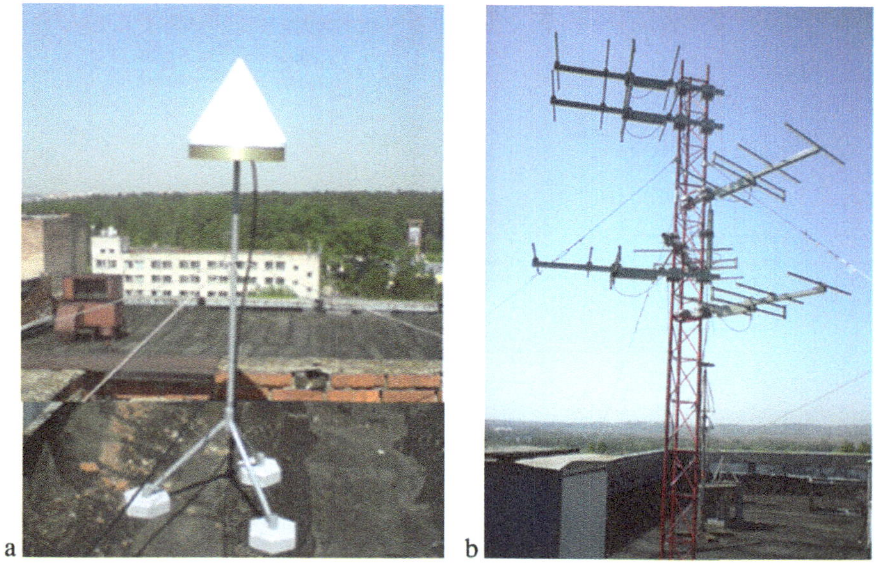

Fig. 3.45 The location of antennas on the roof of the building: **a** SNS antennas, **b** VDB data link (DL) antennas

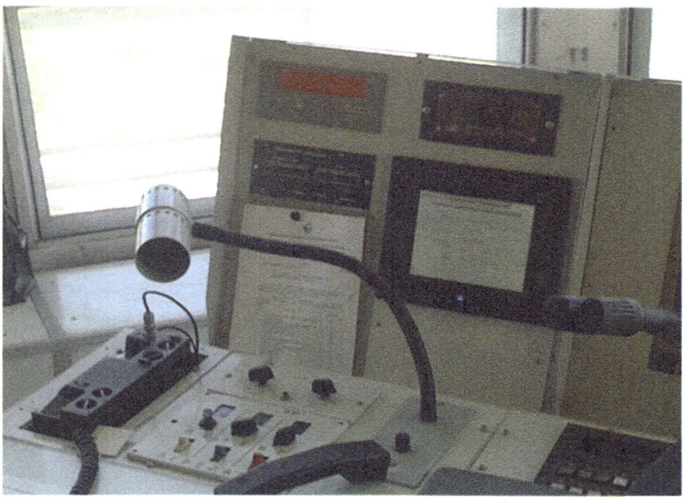

Fig. 3.46 The location of the flight manager console and landing controller console on the control tower

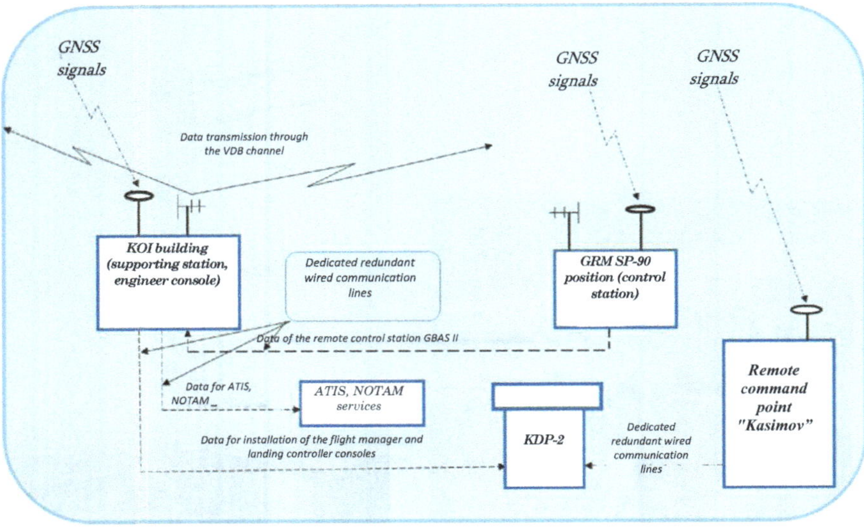

Fig. 3.47 The block diagram of ground lines between GBAS elements

Fig. 3.48 Flying laboratory in flight tests

The display of landing signals and deviations from a set trajectory of the final approach site (FAS) transmitted to the board from GBAS with differential data was carried out in the cockpit on flight-navigation instruments (FD, FNI) from the ACSU-154M and on the indicator.

Figures 3.51, 3.52, 3.53 and 3.54 illustrate the placement of GLS equipment on the board of the Tu-154:

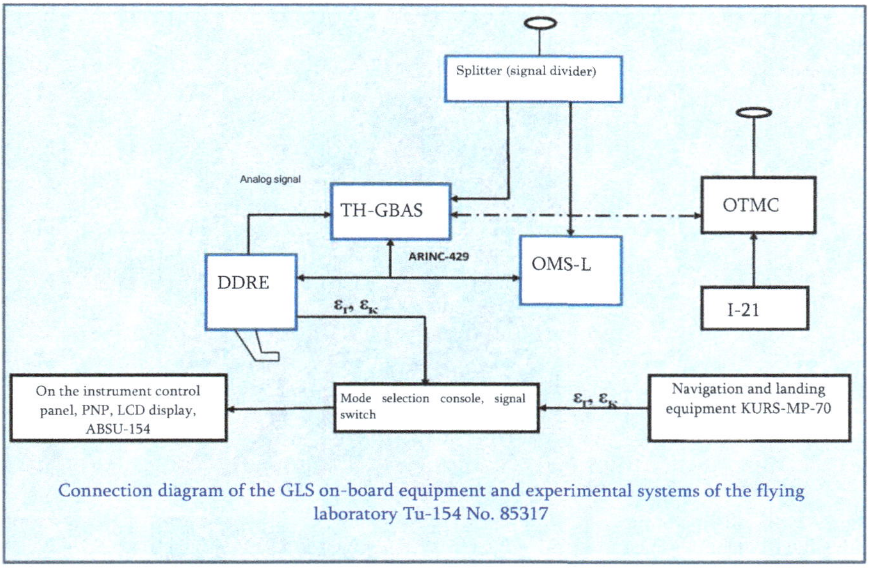

Fig. 3.49 Connection diagram of on-board equipment

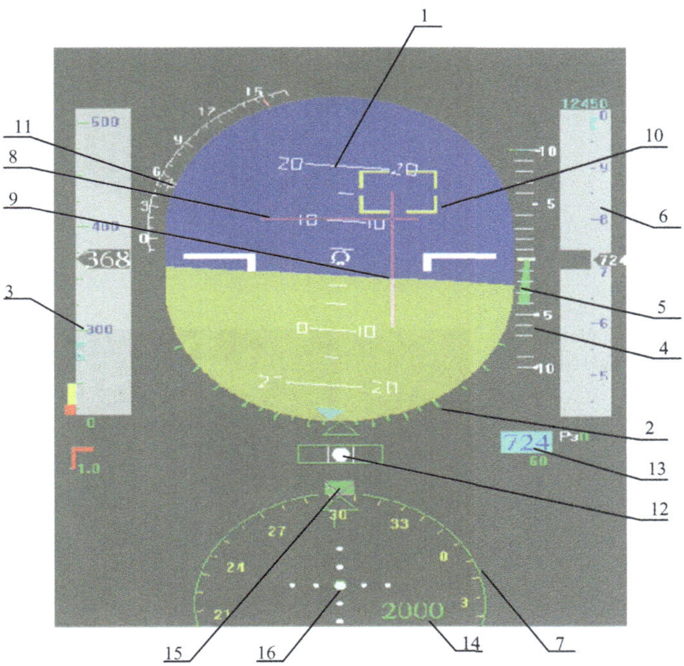

Fig. 3.50 Type of flight information on the indicator

Fig. 3.51 The flying laboratory Tu-154M, testing equipment

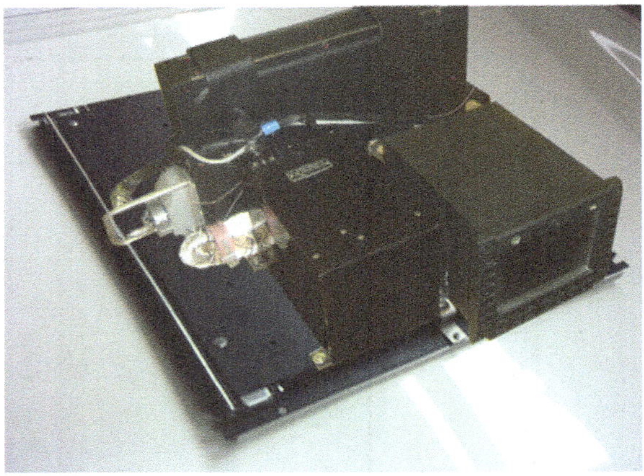

Fig. 3.52 Installation of the on-board GLS and TH-GBAS equipment on a common platform

Fig. 3.53 OMS installation on the rack of the Tu-154 aircraft

Fig. 3.54 View of left pilot instrument display

The list of parameters displayed on the indicator is as follows:

1. Tangage scale [33]
2. Roll scale (bottom, fixed in relation to the device).
3. The scale of the instrument rate (tape, with a number in the window).
4. Vertical rate scale.
5. Vertical speed index.
6. Height scale (tape, with a number in the window).
7. Course scale
8. Director pitch bar.
9. Director roll bar.
10. Square index of course and glide path deviation.
11. Indicator of angle of attack.
12. Slip indicator ("ball").
13. Radio altitude counter.
14. Range counter to the end of a FLS.
15. Index of a set course.
16. Indicator of the angular deviation from the glide path at the course and height in "points".

According to the results of the ground-based airfield and flight tests of GBAS satellite navigation and landing equipment of the II and III categories on the IL-76 and IL-103 laboratory airplanes under conditions of actual interference of the airfield, the act was drawn up (Annex E).

In the process of flight tests, a huge amount of factual material was obtained from navigation measurements, which was processed by statistical methods using the Markov theory of estimation of random processes. The analysis of statistical characteristics showed that the process of determination of the errors of differential corrections is stationary and ergodic.

Determination of the basic statistical characteristics of LMCS stationary processes was based on the following theoretical principles of mathematical statistics.

The ergodic process $\xi(t)$ is a stationary process, the probabilistic characteristics of which can be obtained from a single realization (of a sufficiently long duration) with the probability equal to one.

The stationary process $\xi(t)$ is called ergodic if, with the probability equal to one, all its probabilistic characteristics can be obtained by a single realization of the process of the sufficiently long duration T.

The statistical characteristics of the estimate of an error of differential corrections to the LMCS PR, averaged over time at the annual measurement interval, coincide with the corresponding ensemble averaging results. The analysis of the received posterior data confirms the ergodic nature of the errors of the differential data.

In this case, the probabilistic characteristics of the ergodic process are determined by the time averaging of realizations over the period of time $[0, T]$, after which the parameters are set at which the dispersion of the resulting estimates tends to zero at $T \to \infty$.

The random stationary process $\xi(t)$ has ergodic properties with respect to the assembly average (AA), i.e., [34]

$$M\{\xi(t)\} = m_\xi = \lim_{T\to\infty} \frac{1}{T} \int_0^T \xi(t)dt, \quad (3.30)$$

if and only if the equality is satisfied

$$\lim_{T\to\infty} \frac{2}{T} \int_0^T \left(1 - \frac{\tau}{T}\right) R_\xi(\tau)d\tau = 0, \quad (3.31)$$

where $R_\xi(\tau)$ is the correlation function of the process.

The law (3.31) is confirmed by the calculation based on measurement data on the accuracy of differential corrections (Figs. 3.16a, b, 3.17a, b, 3.18a, b, 3.19a, b, c, 3.20a, b, c).

The random stationary process $\xi(t)$ has ergodic properties with respect to the correlation function, i.e.,

$$R_\xi(\tau) = \lim_{T\to\infty} \frac{1}{T} \int_0^T [\xi(t) - m_\xi][\xi(t+\tau) - m_\xi]dt, \quad (3.32)$$

if and only if the correlation is satisfied

$$\lim_{T\to\infty}\frac{2}{T}\int_0^T\left(1-\frac{\tau}{T}\right)R_\chi(\tau)d\tau = 0, \tag{3.33}$$

where $R_\chi(\tau)$ is the correlation function of the process $\chi(t) = \xi(t)\xi(t+\tau)$.

The random stationary process $\xi(t)$ has ergodic properties with respect to a one-dimensional distribution function $F(x)$, i.e.,

$$F(x) = \underset{T\to\infty}{\mathrm{l.i.m}}\frac{1}{T}\int_0^T\eta(t)dt,\quad \eta(t) = \begin{cases}0,\ \xi(t) > x\\ 1,\ \xi(t)\le x\end{cases}, \tag{3.34}$$

if and only if the equality is satisfied

$$\lim_{T\to\infty}\frac{1}{T}\int_0^T\left(1-\frac{\tau}{T}\right)\left[F(x,x;\tau)-F^2(x)\right]d\tau = 0, \tag{3.35}$$

where $F(x_1, x_2; t)$ is two-dimensional process probability distribution function $\xi(t)$.

The ergodic properties of stationary processes in the statistical evaluation of DC errors with respect to the one-dimensional distribution function F(x) are used in the construction of algorithms for rejection of the excess of signal-to-noise ratio in the presence of noise interference.

The correlations (3.30)–(3.35), written for continuous in time realization of the random process $\xi(t)$, can be represented in an equivalent form, corresponding to the definition of statistical characteristics by values $\xi_i = \xi(t_i)$ at the discrete points in time $t_i = (i-1)\Delta t$, where $1 \le \mathrm{i} \le \mathrm{N}$, $N = [T/\Delta t + 1]$, Δt is the time step.

In particular, for AA we have

$$m_n^* = \frac{1}{n}\sum_{i=1}^{n}\xi_i,\ n\le N, \tag{3.36}$$

and the dispersion of this estimate is determined by the expression

$$D\{m_n^*\} = \frac{D_\xi}{n^2}\left[n + \frac{2}{D_\xi}\sum_{k=1}^{n-1}(n-k)R_\xi(k)\right], \tag{3.37}$$

where $R_\xi(k) = R_\xi(k\Delta t)$, and D_ξ is the true value of the process dispersion ξ.

The analysis of the empirical data on the accuracy of the formation of differential corrections over a long time interval showed that the assembly average of errors is equal to the limit of the integral of the random variable over time, provided that the dispersion of the assembly average tends to zero.

The estimate of the correlation function at $0 \leq k \leq N - 1$ for the sample of the total volume N is [33, 35]

$$R_{\xi}^{*}(k) = \frac{1}{N-k} \sum_{i=1}^{N-k} \left(\xi_i - m_N^*\right)\left(\xi_{i+k} - m_N^*\right), \tag{3.38}$$

and the dispersion of such an estimate is set by the approximate ratio

$$D\left\{R_{\xi}^{*}(k)\right\} \approx \frac{1}{N} \sum_{j=-\infty}^{\infty} \left[R_{\xi}^{2}(j) + R_{\xi}(j+k) + R_{\xi}(j-k)\right]. \tag{3.39}$$

It follows from (3.37) that if $D\left\{m_{\xi}^{*}\right\} \rightarrow 0$ at $n \rightarrow \infty$, then the AA estimate (3.36) is asymptotically effective and consistent.

According to the sample of the maximum possible volume for this experiment N, the dependencies of the correlation function estimates $R_{\xi}^{*}(k), 0 \leq k \leq N - 1$, are calculated and constructed according to the formula (3.38), based on (3.37)—the dependence of the AA estimate m_n^*, $n \leq N$, and in accordance with the expression

$$D_n^* = \frac{1}{n-1} \sum_{i=1}^{n} \left(\xi_i - m_n^*\right)^2, \quad n \leq N, \tag{3.40}$$

unbiased dispersion estimate.

Dependencies of estimates of the assembly m_n^* average and dispersion D_n^*(or RMSD σ_n^*) on the sample length n allow to reveal the nature of the convergence of the estimates to the true values of the parameters. The absence of convergence means that either the sample size is insufficient or the process being analyzed is not stationary. The graph of the estimate of the normalized correlation function

$$r_{\xi}^{*}(k) = R_{\xi}^{*}(k)/D_N^* \tag{3.41}$$

allows to determine approximately the correlation time τ_k, which allows to judge approximately on what time interval on average there is a noticeable correlation between the values of a random process that is essential for the task being solved.

With this in mind, thse sample of the total volume N can be represented in the increments $m = [\tau_k/\Delta t]$, with the result that we obtain a sample of a smaller volume $N_H = [(N-1)/m] + 1$, but consisting of approximately independent counts of the correlated process under study. Such a procedure usually allows to find more accurate estimates of distribution points and more easily determine the accuracy of estimates for a given sample size.

In particular, to assess the AA

$$m_{N_H}^* = \frac{1}{N_H}\sum_{j=1}^{N_H}\xi[1+(j-1)m] = \frac{1}{N_H}\sum_{j=1}^{N_H}\xi_1[j], \tag{3.42}$$

where $\xi_1[j]$ are the elements of the singled out sample, instead of (3.37) we get

$$D\{m_{N_H}^*\} = D_\xi/N_H. \tag{3.43}$$

Similarly, for an unbiased estimate of the process dispersion we have

$$D_{N_H}^* = \frac{1}{N_H - 1}\sum_{j=1}^{N_H}\left(\xi_1[j] - m_{N_H}^*\right)^2, \tag{3.44}$$

and the dispersion of this estimate is determined by the expression

$$D\{D_{N_H}^*\} = \frac{N_H}{(N_H-1)^2}\left[\mu_4 - D_\xi^2 - \frac{2\left(\mu_4 - 2D_\xi^2\right)}{N_H} + \frac{\mu_4 - 3D_\xi^2}{N_H^2}\right], \tag{3.45}$$

where μ_4 is the true value of the central moment of the fourth order. For the Gaussian process, for which $\mu_4 = 3D_\xi^2$, it follows from (3.45) that

$$D\{D_{N_H}^*\} = \frac{2D_\xi^2}{N_H - 1}. \tag{3.46}$$

The calculation of estimates of the assembly average and the standard deviation in the post-processing begins with the construction of the process implementation schedule. With some experience, the type of implementation of a random process makes it possible to judge whether this process can be considered (at least approximately) stationary.

Then, the calculation of the estimates of the assembly average and the standard deviation is made by the recurrence relations, which follow from (3.36) and (3.40):

$$m_n^* = \frac{1}{n}\sum_{i=1}^{n}\xi_i = \frac{m_{n-1}^*(n-1) + \xi_n}{n}, \ 1 \le n \le N, \tag{3.47}$$

$$D_{2,n}^* = \frac{1}{n-1}\sum_{i=1}^{n}\xi_i^2 = \frac{D_{2,n-1}^*(n-2) + \xi_n^2}{n-1}, \ 2 \le n \le N, \tag{3.48}$$

$$D_n^* = D_{2,n}^* - \left(m_n^*\right)^2, \quad 2 \le n \le N, \tag{3.49}$$

$$\sigma_n^* = \sqrt{D_n^*}, \quad 2 \le n \le N. \tag{3.50}$$

As already noted, the graphs of the estimates of AA m_n^* and RMSD σ_n^* from the sample length n allow to reveal the nature of the convergence of the estimates to the true values. The results are presented in Figs. 3.23 and 3.24.

Figure 3.55 presents the general post-processing scheme for standard requests.

As a result of the work done on processing of a large array of navigation measurements, the structure and sequence of the post-processing process of data received from the LMCS test sample with the aim of obtaining real-time data using mathematical software using the database were determined (Fig. 3.56).

The general scheme for the implementation of post-processing of monitoring data includes seven steps [36, 37]:

1. The network access application asks the user for the method of summarization of data, the source (location) and the period (time).
2. The post-processing service forms a structured request to the database storing the loaded data.

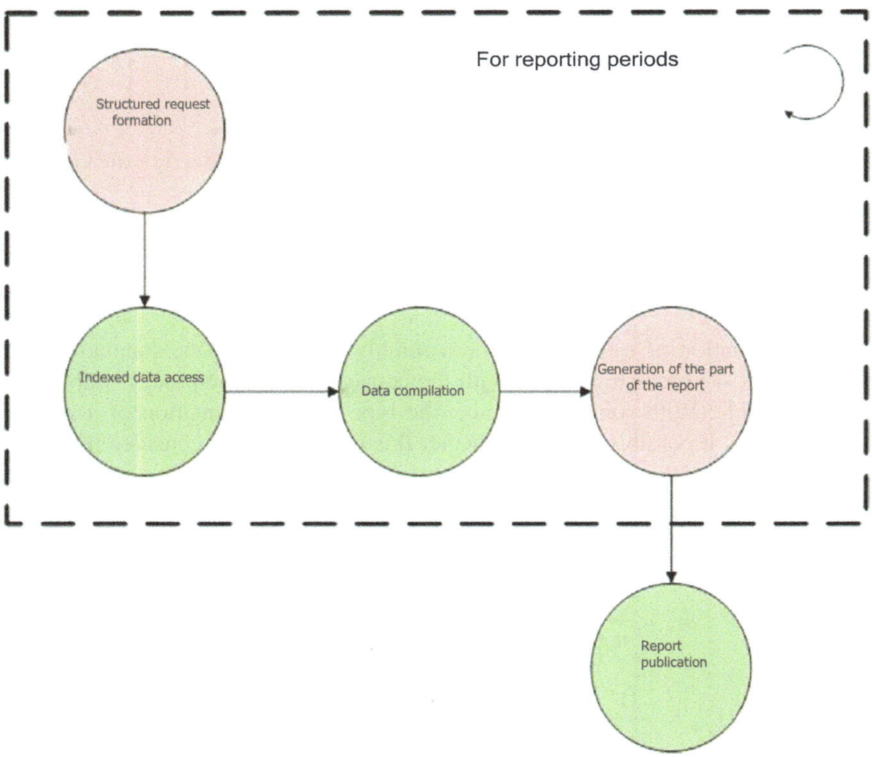

Fig. 3.55 The general post-processing scheme for standard queries

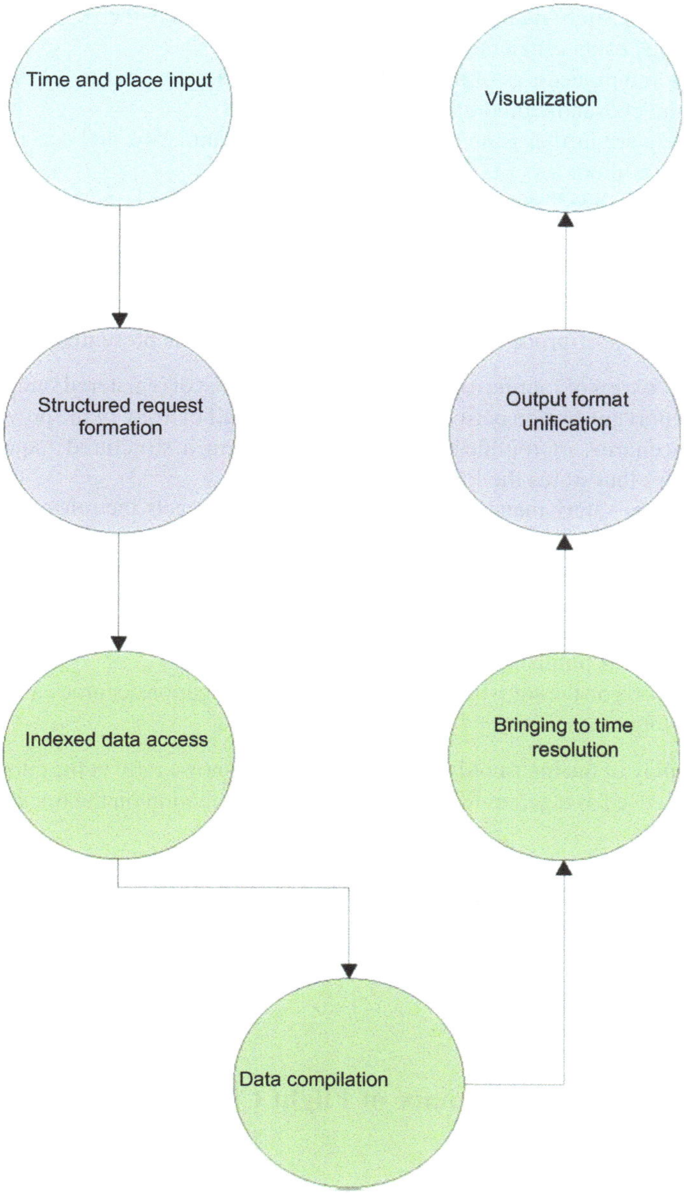

Fig. 3.56 General post-processing scheme for user requests

3. The traffic safety management system (TSMS) extracts the relevant data from the tables using efficient index access.
4. The stored procedures of the TSMS perform data transformation in accordance with the chosen method of generalization.
5. The data are further generalized to reduce the granularity, and accordingly, the volume without loss of significant information.
6. The generalized data are unified and transferred to the network access application.
7. The access application performs data visualization in accordance with the developed data presentation formats.

The general post-processing scheme for user requests is presented in Fig. 3.48.

1. The scenarios of routine report generation form a list of registered sample models and report generation periods (first and second half of a day of the previous day).
2. The scenarios of routine report generation form a structured request to the database that stores the loaded data.
3. The traffic safety management system (TSMS) extracts the relevant data from the tables using efficient index access.
4. The stored procedures of the TSMS perform data transformation in accordance with the chosen method of generalization.
5. The summarized data is added to the generated report.
6. The report is published for review by users.
7. The reports on the statistical characteristics of navigation accuracy are generated according to a sample.

The display of the measured value of the signal-to-noise ratio at the current elevation angle is used as a generalization of the basic information on the tracking quality for satellites.

Under normal conditions, the ratio increases uniformly with increase of elevation angle and at least 6–8 satellites with the signal-to-noise ratio of more than 45 dB * Hz should be observed.

The following statistical results were obtained in the course of navigation data processing during flight tests (Table 3.5).

3.4 Metrological Assurance of Flight Checks of LMCS (GBAS)

For the purpose of metrological assessment of satellite landing parameters and the accuracy of the differential mode, the ground and flight test complex is proposed which was evaluated by the All-Russian Scientific Research Institute of Physical-engineering and Radiotechnical Metrology (VNIIFTRI) and received the Certificate of measuring device type of the Federal Agency on Technical Regulation and Metrology.

Table 3.5. The results obtained during the navigation data processing during flight tests

Mode number	PEE, 95%, (m)		FTE, 95%, (m)		TSE, 95%, (m)		Mode time (s)	Piloting mode
	Hor	Vert	Hor	Vert	Hor	Vert		
Mode 1	0.8	0.5	6.0	4.3	6.0	4.3	100	Automatic
Mode 2	0.7	1.4	5.4	4.8	5.4	5.0	102	Automatic
Mode 3	1.4	0.3	4.2	4.5	4.4	4.5	108	Automatic
Mode 4	0.9	0.9	5.2	3.2	5.3	3.3	96	Automatic
Mode 5	2.2	1.9	4.0	2.2	4.6	2.9	111	Automatic
Mode 6	1.8	1.8	4.3	3.5	4.7	4.5	113	Automatic
Mode 7	1.5	0.8	4.8	3.9	5.0	4.0	105	Director
Mode 8	1.4	1.3	4.1	4.2	4.7	4.4	100	Director
Mode 9	0.8	1.0	2.4	4.1	2.5	4.2	110	Automatic
Mode 10	1.3	0.9	5.3	2.1	5.5	2.3	112	Automatic
Mode 11	0.8	1.3	4.3	3.2	4.4	3.5	70	Director
Mode 12	1.2	1.5	5.2	2.8	5.3	3.2	75	Director
Mode 13	0.7	2.0	6.0	1.9	6.0	2.8	72	Director
Mode 14	1.4	0.4	5.2	4.8	5.4	4.8	66	Director
Mode 15	0.5	0.6	5.9	3.0	5.9	3.1	73	Director
Mode 16	1.4	1.3	4.3	3.1	4.5	3.4	70	Director
Mode 17	1.8	1.9	3.0	4.6	3.5	5.0	75	Director
Mode 18	0.4	0.9	4.4	2.2	4.4	2.4	72	Director
Mode 19	2.0	1.9	4.5	4.2	4.9	4.7	65	Director
Mode 20	0.5	1.2	5.5	3.9	5.5	4.5	63	Director

The ground and flight test complex (hereinafter—GFTC) is designed to determine own geodetic coordinates of equipment locations, the coordinates of local monitoring and control stations (LMCS) and to quantify the parameters of GNSS signals (GLONASS/GPS/Galileo) and their space (SBAS) and ground (GBAS) functional extensions and to test ground-based radio flight support equipment (RFS), airfields lighting equipment (LE) and zonal navigation characteristics based on the determination of GFTC coordinates.

GFTC is the hardware/software complex for the quantitative assessment of the parameters of signals generated and transmitted by LMCS and SBAS, which are used as the main means of navigation on the flight route, in the area of airfields, including to ensure the approach, during their testing and operation.

GFTC provides reception and processing with the frequency of at least 2 Hz of the navigation signals of GNSS satellites: GLONASS in the frequency ranges (1602.56–1615.50) MHz (L1), (1246.00–1256.50) MHz (L2) and 1202.025 MHz (L3); GPS NAVSTAR at frequencies of 1575.42 MHz (L1), 1227.60 MHz (L2) and 1176.45 MHz (L5); optional is Galileo in the frequency bands 1559–1592 MHz (E1) and (1164–1215) MHz (E5). GFTC also provides reception and processing of LMCS correction information and/or SBAS functional extensions (WAAS, EGNOS, MSAS, SDCM). It is used autonomously or as part of an automated flight control system (AFCS) of an aircraft/helicopter laboratory, or a specially designated aircraft.

Based on the measurement results, GFTC quantifies the following aircraft navigation parameters: absolute positioning error in the GNSS differential mode; the strength of the radio field created by the LMCS and/or SBAS functional extensions; the conformity of the differential data received and measured by reception equipment (DDRE) and the assigned LCSS and/or functional extensions of parameters; and the deviation from the desired track (DT) of an aircraft. When processing the received measurement information, the presence (or absence) of interference in the frequency range of satellite navigation signals is determined.

According to the measurement data, GFTC calculates the coordinates of the location in the autonomous and/or differential modes, compares the calculated coordinates with the established coordinates of the phase center of the antenna in the PZ 90.02 or WGS-84 coordinate systems and calculates the deviations. If the deviations do not exceed the v tolerances for the accuracy of navigation of an aircraft at various stages of its flight (typical flight operations—flight along the route, maneuvering in the airfield area, landing approach, etc.), the complex generates and displays information with a sign of admissibility of a particular operation on the display screen; if deviations exceed the set tolerance, then information is transmitted with a sign of inadmissibility of the operation. GFTC determines the values of the discrepancy of coordinates measured in the differential mode and reference geodetic coordinates in real time and in the post-processing mode.

The GFTC equipment, except for antenna-feeder devices (AFD), is made in a single structural module, enclosed in a protective plastic container. The module includes the following blocks: satellite receiver navigation unit (SRU); computing and switching system unit (CSSU) with a hard real-time operating system, including a touch-screen display; unit converters; DDRE block—calibrated VDB receiver. The external devices in the GFTC are the AFD: a geodetic receiving antenna, which provides the operation of the SRU; VHF-band antenna ensuring the operation of the DDRE; antenna cables.

On the GFTC operating panel there are on/off keys, a touch screen display and connectors for connecting antennas, an external computer and a local area network. There are additional service connectors on the panel for input and output of analog and digital signals of the ARINC 429 standard according to GOST 18,977–79. The operation of GFTC is performed using the touch panel of the built-in computer of CSSU.

GFTC accepts corrective information of functional extension in the format compliant with SARPs on GNSS of ICAO (Annex 10, Volume I). GFTC also provides the user with the MIL-STD-1553B interface (optional). The appearance of the GFTC is shown in Fig. 3.57.

Under normal operating conditions GFTC operates automatically and does not require external control actions after the initial setup. GFTC provides measurements of the radio field strength in the frequency range of 108–118 MHz with the channel spacing of 25 kHz with horizontal or elliptical polarization of the VHF antenna. GFTC provides registration of data accumulated during ground or flight verification, their storage and the possibility of removal to external media for further processing. GFTC determines the value of the discrepancy of the coordinates measured in the differential mode of the SNS and geodetic or reference coordinates in real time and in post-processing.

The designated parameters for GFTC are:

- The source of differential corrections;
- The GNSS ICAO SARP's format;
- The geodetic coordinates of the phase centers of the antennas of the SNS in the coordinate system PZ-90.02/WGS-84;
- The files of the transferred information of functional extensions to GNSS.

In accordance with the Order of the Head of the Federal Air Transport Agency on the use of ground and flight test complex (GFTC), FSUE "State ATM Corporation", together with LLC RDPF Spektr and the branch Aerocontrol, carried out the field tests of GFTC equipment with an automated flight control system (AFCS), installed on the board of the aircraft laboratory "BeechcraftKingAir 350i", the board RA № 02,800.

Test sites: airports of Rostov-on-Don, Sochi, Anapa, Mineralnye Vody, Krasnodar.

Fig. 3.57 The appearance of GFTC

The following results were obtained during the tests (Fig. 3.58):

The ergonomic characteristics of the GFTC allow the operator to assess visually the LMCS characteristics obtained during a flight. Connecting the GFTC to the technological output of an AC course antenna ensures stable reception of the radio signal in the LMCS transmission range. SNS antennas of GFTC provide a steady reception of the signals of the GLONASS and GPS navigation groups (Fig. 3.59).

An error in the GLONASS/GPS coordinates obtained by the GFTC aboard an aircraft in the differential mode according to the data of LMCS-A-2000 was not more than 1.5 m in any direction (ICAO requirements: 6–10 m). The RMSD of an error of the GLONASS/GPS coordinates of externally trajectory measurements of the GFTC in any direction did not exceed 0.5 m (Figs. 3.60 and 3.61).

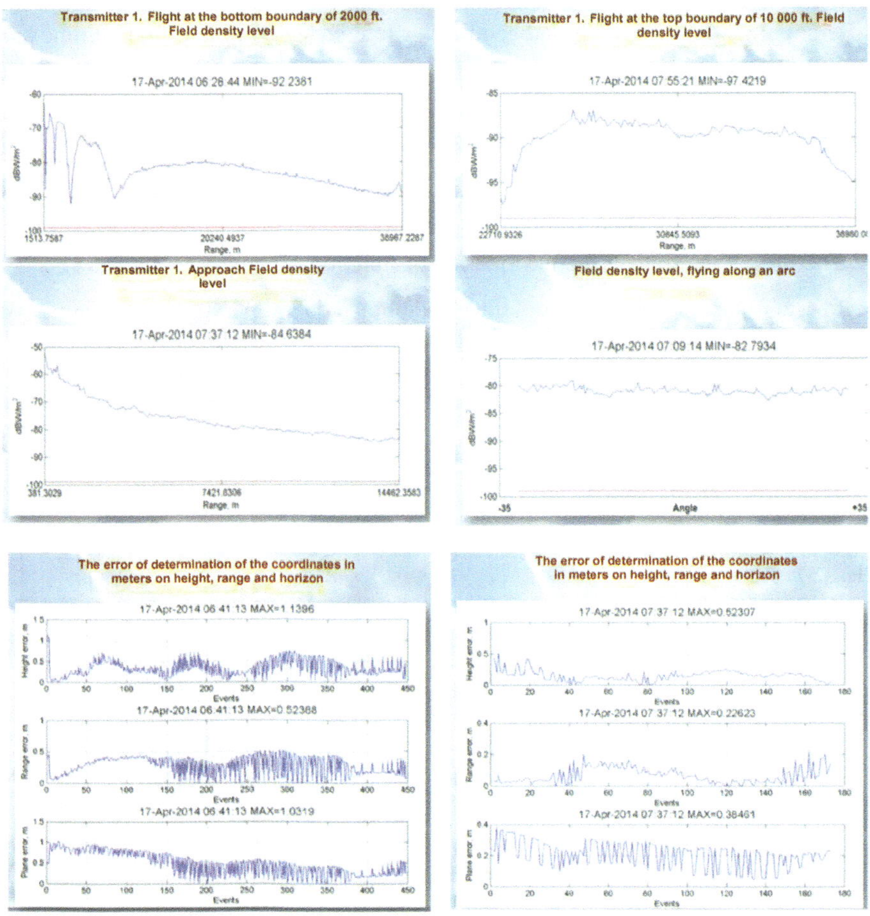

Fig. 3.58 The results of tests of GFTC

a

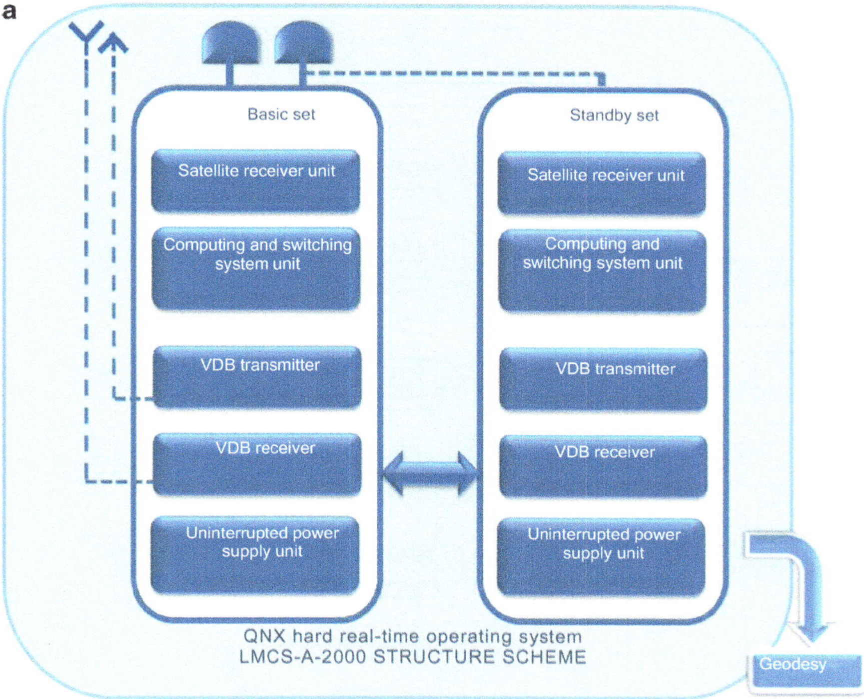

Fig. 3.59 **a** The structural diagram of the satellite navigation and landing system of LMCS-A-2000 **b** The functional scheme of the satellite navigation and landing system of LMCS-A-2000

GFTC provides registration, removal and storage of received digital data. Radio reception of digital data transmitted from the LMCS-A-2000 and their comparison with the data stored in the GFTC is provided for all communications.

Radio signal reception in GFTC is provided by certified differential data reception equipment (DDRE), which complies with the requirements of the "Methodical recommendations for flight tests of ground-based radio equipment for flight operations, aviation telecommunications and airfield lighting systems of civil aviation", approved by Order of the Ministry of Transport of the Russian Federation from 29.06.2012 MS-82-r (Annex B).

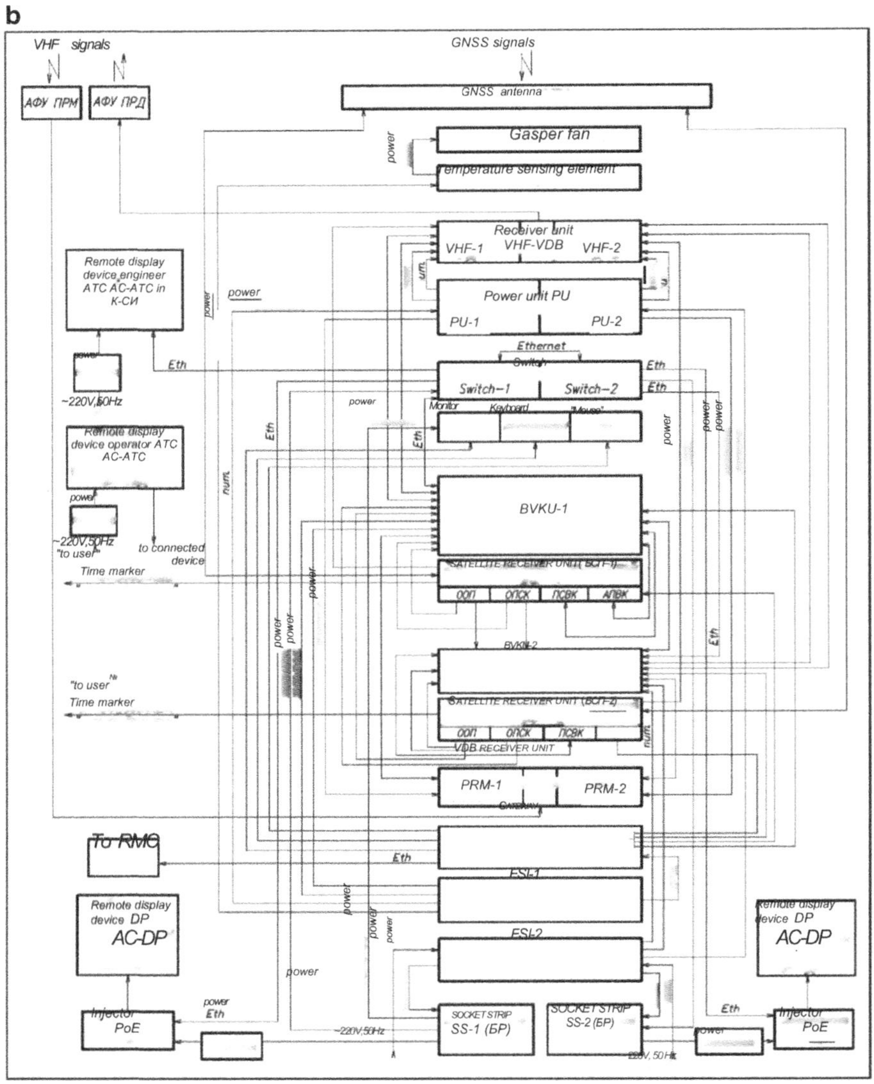

Fig. 3.59 (continued)

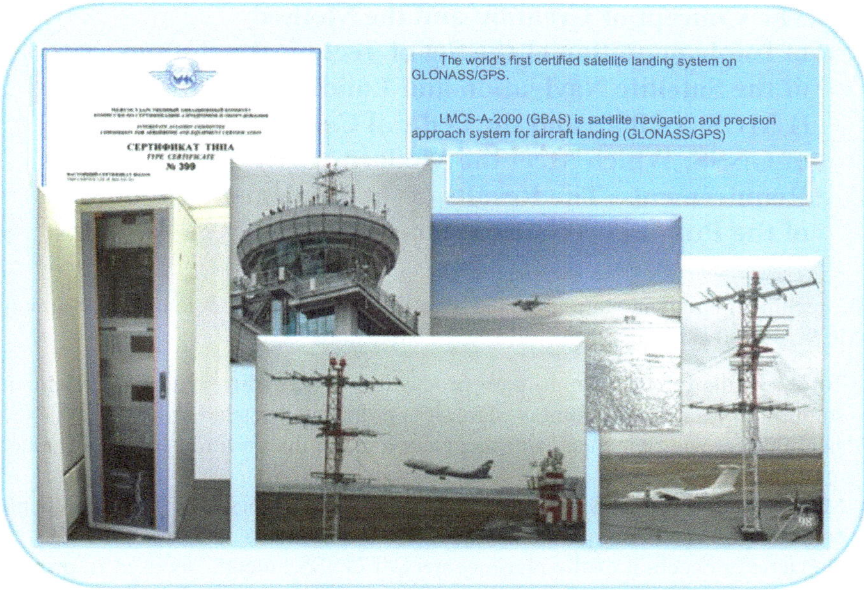

Fig. 3.60 The satellite navigation and landing system of LMCS-A-2000 (implementation)

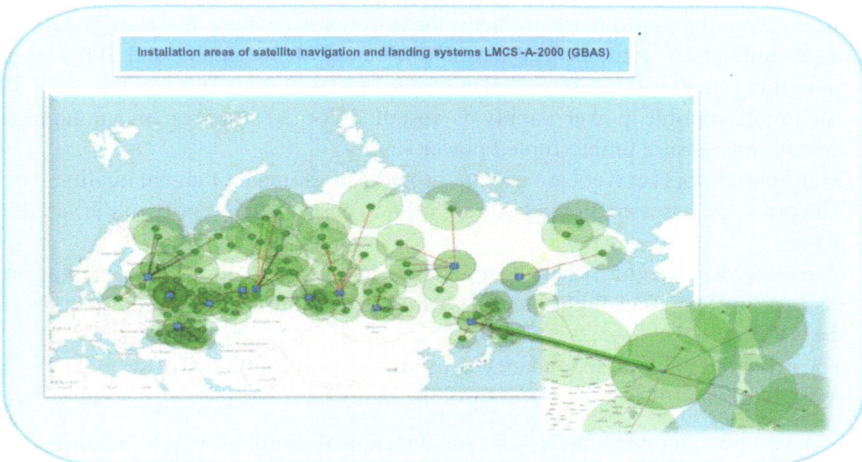

Fig. 3.61 Installation areas of satellite navigation and landing systems of LMCS-A-2000

3.5 The Concept of Creation and the Method of Implementation of the Set of Technical Solutions of the Satellite Navigation and Landing System (LMCS-A-2000) and the Federal Centre for Monitoring of GNSS, Which Meet the Modern ICAO Requirements. The Results of the Implementation of the Proposed Technical Solutions

The technical implementation of the satellite navigation and landing system of LMCS-A-2000 is based on the following conclusions [37]:

- To assess the effect of multi-path propagation of a radio signal in a technical solution of a satellite landing system, it is proposed to install at least two GNSS satellite receiving antennas and receivers located in conditions of uncorrelated multi-path;
- Multi-constellation GPS + GLONASS + Galileo receivers are installed to ensure continuity of service requirements;
- Geodetic connection of the phase centers of GNSS antennas is used to provide differential mode;
- A remote control system is installed to provide remote control of the differential mode;
- To estimate the ionospheric gradient, the difference between the code and phase measurements is used, as well as the difference of 30 second and 100 second smoothing intervals of differential corrections;
- An uninterruptible power supply is installed for 30 minutes of autonomous operation to ensure uninterrupted power supply;
- Multi-correlator receivers are used to provide an estimate of signal quality;
- The proposed assessment methods are used to ensure the integrity of the satellite signal;
- Consoles of the landing dispatcher and the engineer, as well as a removable hard disk for investigating flight incidents, are installed to meet the requirements of the ATM;
- The principle of "hot" redundancy is used to ensure continuity of service;
- A VDB receiver is installed to control the radio.

The method of implementation of a set of technical solutions, that is, the structural and functional diagrams of the satellite navigation and landing system of LMCS-A-2000 are shown in Fig. 3.59a, b.

More than 100 major federal airports are equipped with the LMCS-A-2000 (GBAS) system

The Federal Air Transport Agency issued the Order No. AU-142-r "On the acceptance for equipping the ground local monitoring and control station of the GLONASS/GPS combined navigation satellite system LMCS-A-2000" at civil aviation airports (Annex B).

The obtained results of accuracy in determination of the coordinates in the differential mode using the GFTC flight test complex described in Sect. 3.4 following the results of tests in Sheremetyevo Airport are shown in Figs. 3.62, 3.63 and 3.64.

An analysis of the presented results of measurements of the accuracy of coordinates in the differential mode shows that the accuracy required by the ICAO standards, even with the categorized approach (including vertical guidance) is provided by the proposed method of implementation of the differential mode. The RMSD error of determination of coordinates was 0.06 m, with the ICAO requirements of 4 m.

The results of flight tests by the FSUE "State ATM Corporation" aircraft laboratory when commissioning the satellite landing system on Iturup Island (the Kuril Ridge) are shown in Figs. 3.65, 3.66, 3.67 and 3.68.

All installed stations of satellite navigation and landing of LMCS-A-2000 for a long period of time confirmed their compliance with the requirements of ICAO standards for GBAS and passed flight tests.

In the Annexes I, H, J and G, the flight test reports of LMCS-A-2000 when equipment was commissioned at the four largest airports in the Russian Federation (Vnukovo, Domodedovo, Pulkovo and Sheremetyevo) are presented.

Information on the implementation of the satellite navigation and landing equipment from the FSUE "State ATM Corporation" is given in the Annex K.

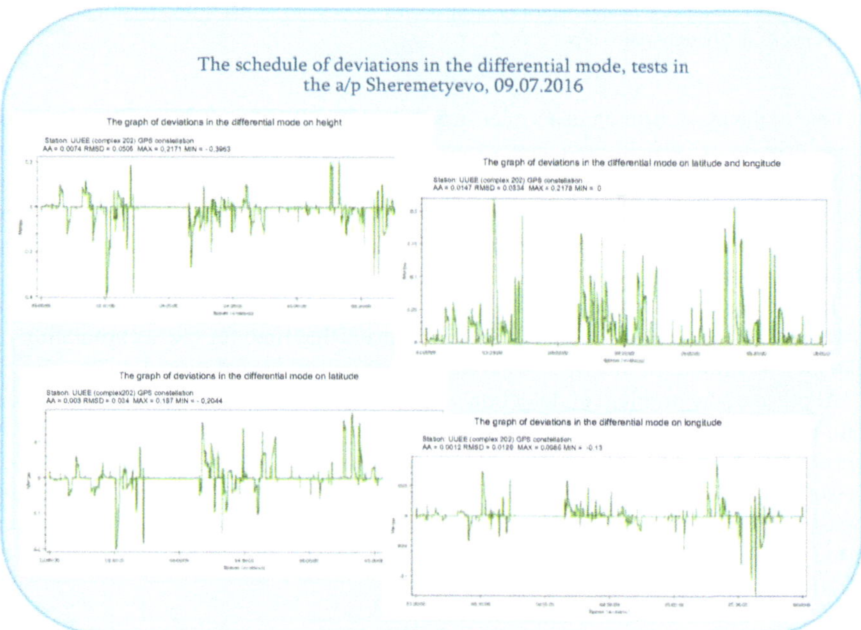

Fig. 3.62 The graphs of deviations of coordinates of an aircraft in the differential mode during flight checks in Sheremetyevo a/p

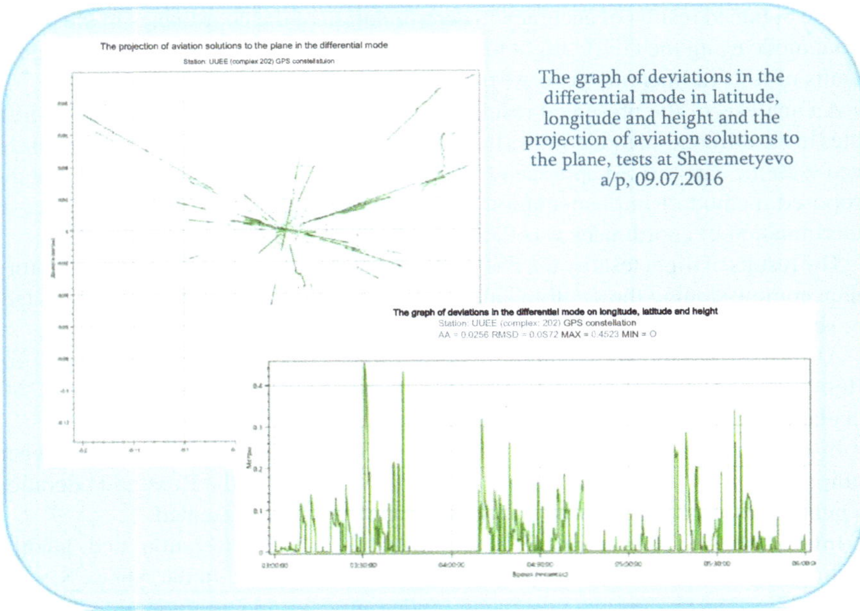

Fig. 3.63 The graphs of deviations of coordinates of an aircraft in the differential mode during flight checks in Sheremetyevo a/p

One of the most difficult tasks of ensuring the integrity of GNSS signals according to the criteria of vertical guidance accuracy and alarm response time is monitoring of the gradient of the ionospheric storm during space and, in particular, solar activity and magnetic storms. The sharp increase in the number of free charged particles in the ionospheric layer of the earth leads to a significant delay in the passage of the electromagnetic GNSS signal to the user's antennas.

The time allotted by ICAO standards for detection of the anomalous effect of a disturbed ionosphere on a signal is 3 s. This means that time for the accumulation of statistics is limited. In fact, the problem must be solved in real time.

At present, the problem of detection of an ionospheric storm in the world is solved with the use of the models Conus and Plasma Buble (for the territories of the USA and Japan). However, the use of these models leads to impairment in the parameters of service continuity, since there is a high probability of rejection of a reliable measurement using a model. In addition, the behavior of ionospheric disturbances varies in different regions. Therefore, there is a need to measure the ion gradient in real time.

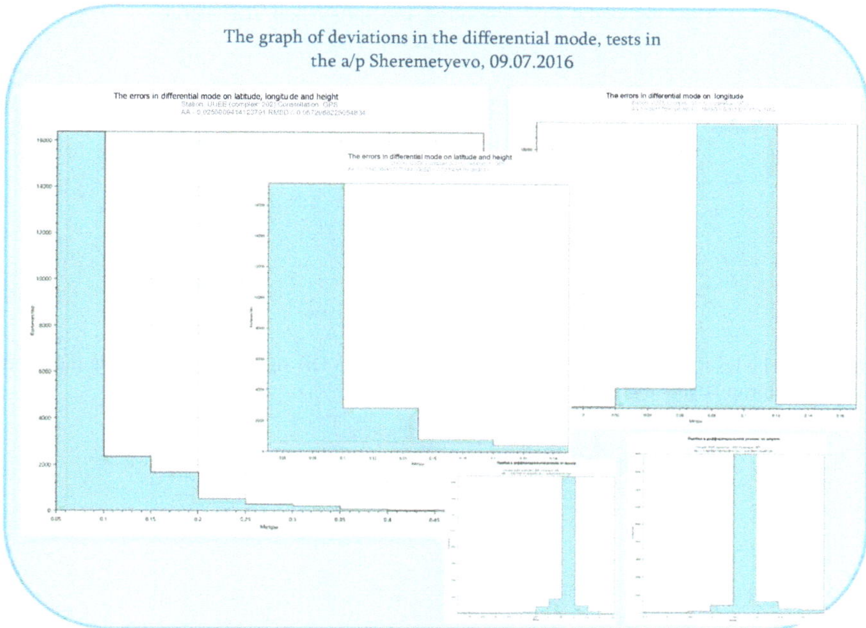

Fig. 3.64 The graph of deviations in the differential mode

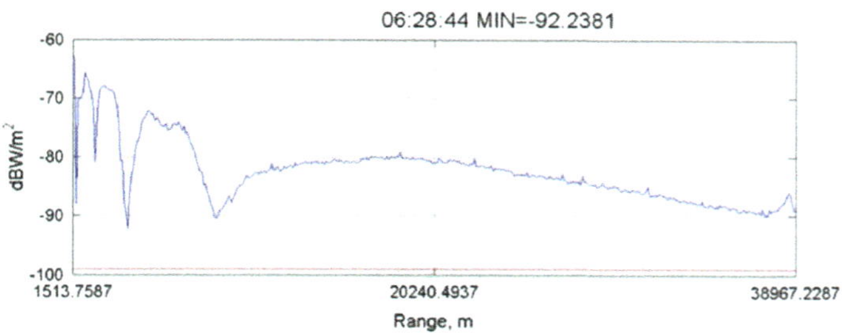

Fig. 3.65 The level of the field strength of the transmitted VDB signal: the flight along the lower boundary (Iturup Island, the Kuril Ridge)

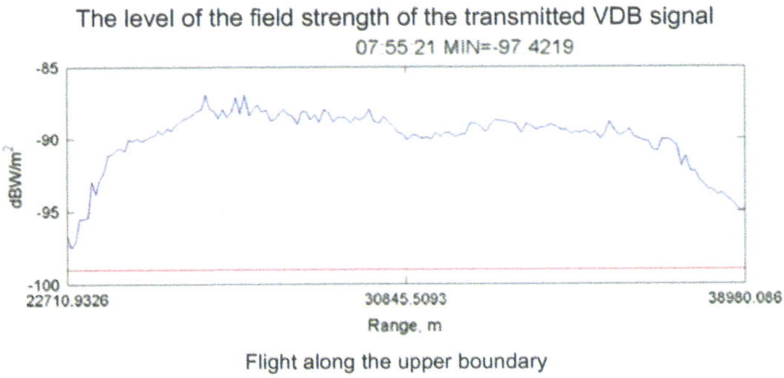

Fig. 3.66 The level of the field strength of the transmitted VDB signal: the flight along the upper boundary (Iturup Island, the Kuril Ridge)

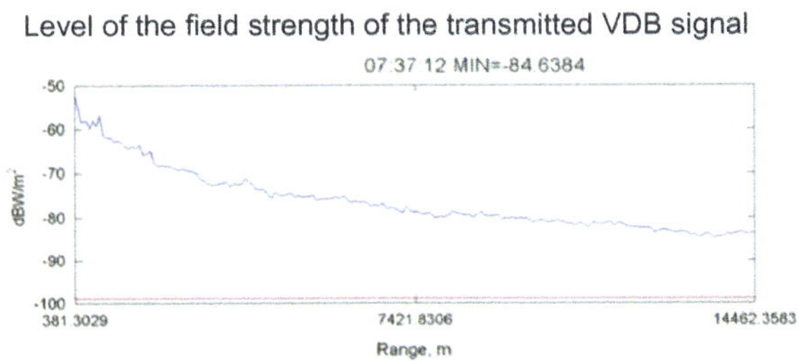

Fig. 3.67 The level of the field strength of the transmitted VDB signal: the flight along the glide path (Iturup Island, the Kuril Ridge)

According to the results of the development of the Federal GNSS Monitoring Centre (the complex automated system for collecting and communicating to the aviation users in the airspace of the Russian Federation of information on the status of orbital groups of the global navigation satellite system and functional extensions CAS CCIM), the acceptance tests were carried out and the Monitoring Centre was commissioned.

To meet the ICAO requirements for informing aircraft crews of the predicted state of GNSS, timely publication of NOTAM of GNSS unavailability, and also recording information on the state of GNSS signals for investigating flight incidents, the analysis of regulatory documents on the need for a GNSS monitoring system for civil aviation is carried out (Annex 10, Volume 1 of SARP's (standards) on ICAO GNSS, Annex 15 SARP's ICAO "Aeronautical Information Services", Doc 9613. Performance Based Navigation (PBN), Volume II, Doc 8071. "Guidance on the tests of radionavigation means", Volume II. "Tests of satellite radio navigation systems", Doc 9849. Instructions for GNSS operation) and the method for the technical implementation of the Federal GNSS Monitoring Centre are proposed.

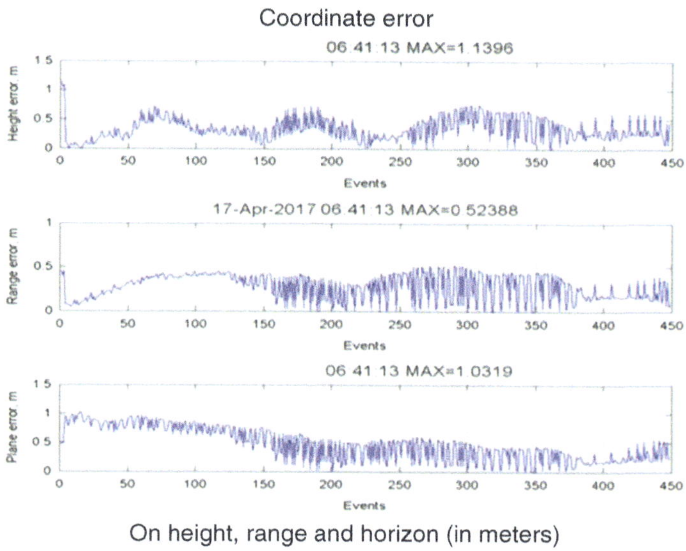

Fig. 3.68 The error of determination of the coordinates on height, range and horizon (in meters) (Iturup Island, Kuril Ridge)

The main requirements of ICAO for the need to monitor GNSS signals (Annex 10, Volume 1 of SARP's (standards) on ICAO GNSS) are:

2.4.3 Registration and storing of received GNSS information

2.1.4.2 The state permitting flight operations using GNSS should include registration of the received GNSS information connected with these flights.

Note 1. The obtained and registered information is intended mainly for use in the investigation of aviation accidents. The obtained GNSS information can also be used to confirm readiness, integrity, continuity and accuracy in the framework of approved operations.

1.1.4.3 The obtained information should be stored for at least 14 days. When investigating accidents, the registered information should be stored for more than 14 days until it is clear that it is no longer needed.

11.1. Registration of GNSS information for the investigation of aircraft accidents is required in the ground-based functional extension system to GNSS and in the corresponding GNSS system used in the operation. The recorded GNSS information depends on the specific system of functional extension and type of operation. The navigation data transmitted to users within the service area must be recorded at representative service area locations.

11.2. The GNSS service should not affect the information recording system. Information for a particular GNSS system should be recorded continuously with the frequency of 1 Hz for the ability to restore the indication of time, aircraft location and rate.

11.3. By the main GNSS systems, it is necessary to record the monitored parameters for all visible satellites:

(a) The quotient “density of the signal on the carrier—noise density” of the observed range-measuring sources;
(b) Carrier phase measurements of observable range-measuring sources and pseudo range code;
(c) Navigation data transmitted by range-measuring sources in the field of view;
(d) The relevant recorded receiver status information.

The set of technical solutions presented by the complex automated system of monitoring and communicating of information to aviation users allowed to meet the above-mentioned ICAO requirements in the Russian Federation.

The GNSS Monitoring System allows the collection, processing and storage of radio monitoring information of the GLONASS and GPS satellite constellations in the airspace of the Russian Federation, visualization of the received data and communicating to aircraft users of information on the admissibility/prohibition of aircraft operations using data from satellite navigation systems, both in autonomous and differential mode of SNS on-board equipment in accordance with the requirements of ICAO.

Figure 3.69a, b represent the functional and structural diagrams of the GNSS Monitoring Centre (CAS CCIM system).

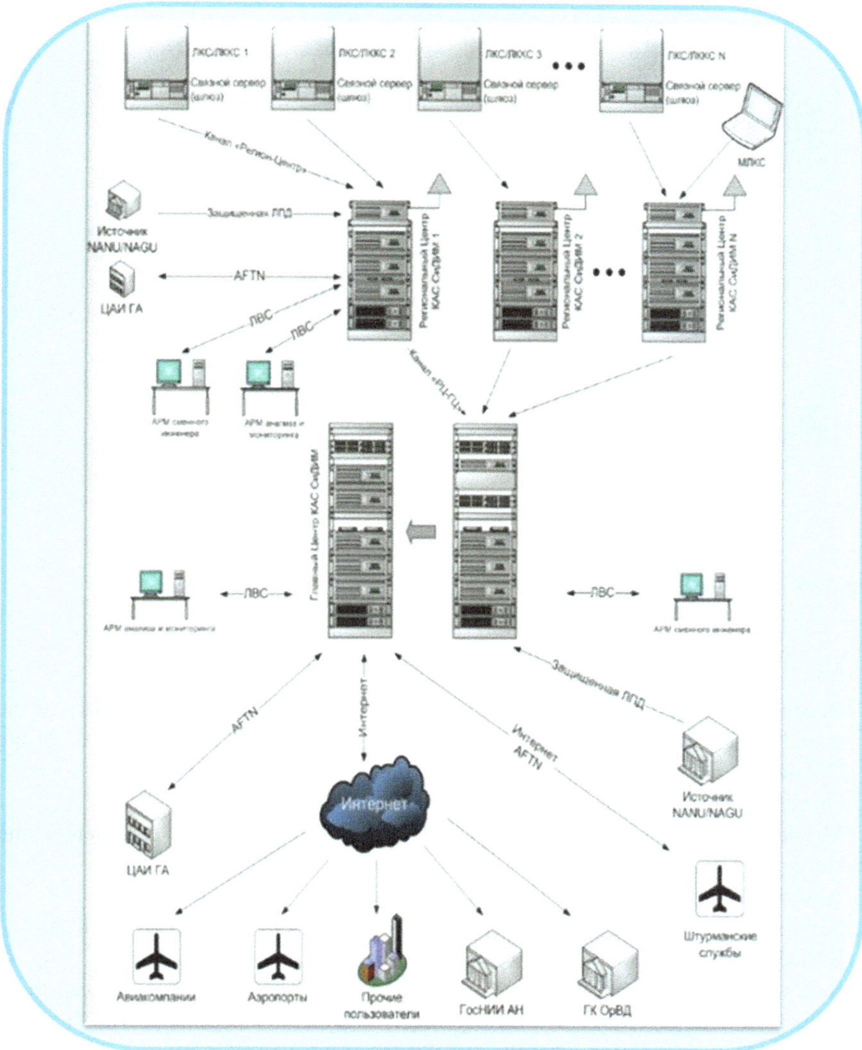

Fig. 3.69 **a** The structural scheme of the GNSS Monitoring Centre **b** The functional scheme of the GNSS Monitoring Centre

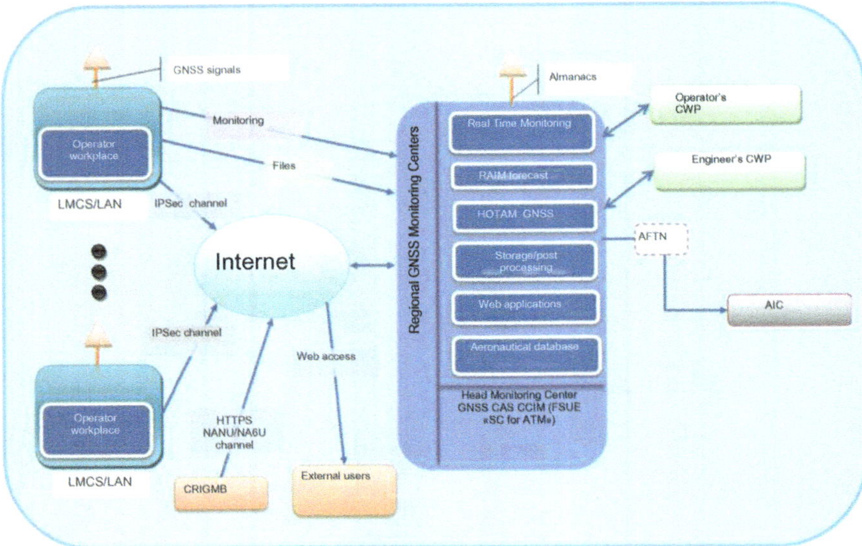

Fig. 3.69 (continued)

The functions of the GNSS monitoring system include:

- Reception of messages from LMCS (MLMCS, LMCS) about the quality of the signals of satellite navigation systems GLONASS/GPS and from the IAC of the Central Research Institute of Machine Building (TsNIIMASH),
- On planned input/output of group satellites (NANU, NAGU),
- Monitoring and registration of all GPS, GLONASS, Galileo parameters and storage for at least 30 days,
- Automatic generation of NOTAM,
- Sending of projects of NOTAM on the unavailability of the RAIM function,
- RAIM availability forecast,
- Recording and storing of GNSS parameters for a flight accident investigation,
- Forecasting the availability of GNSS for performing ROUTE, TERMINAL, NPA operations at destination airports.

Figure 3.70 presents the method for implementation of the GNSS signal monitoring system and the practical results of the work of the Federal Monitoring Centre.

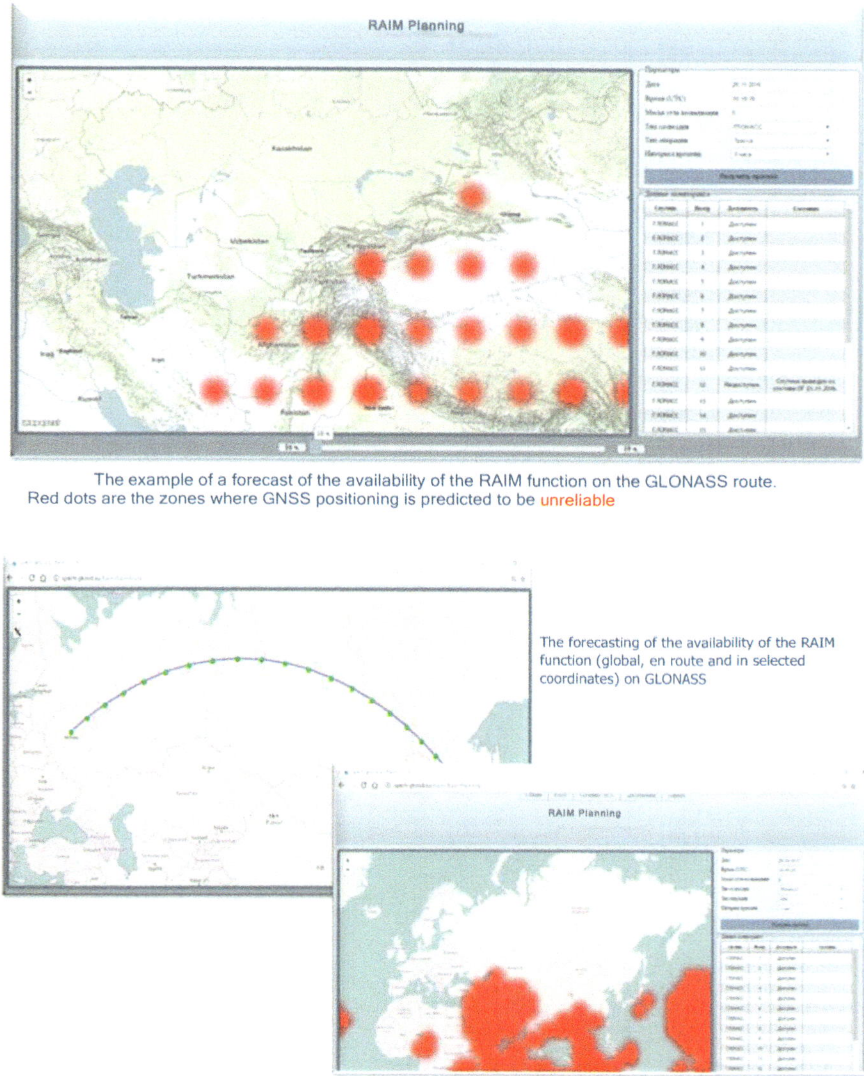

Fig. 3.70 Practical results of the work of the Federal GNSS Monitoring Centre and the way of implementation of the GNSS signal monitoring system in the Russian Federation

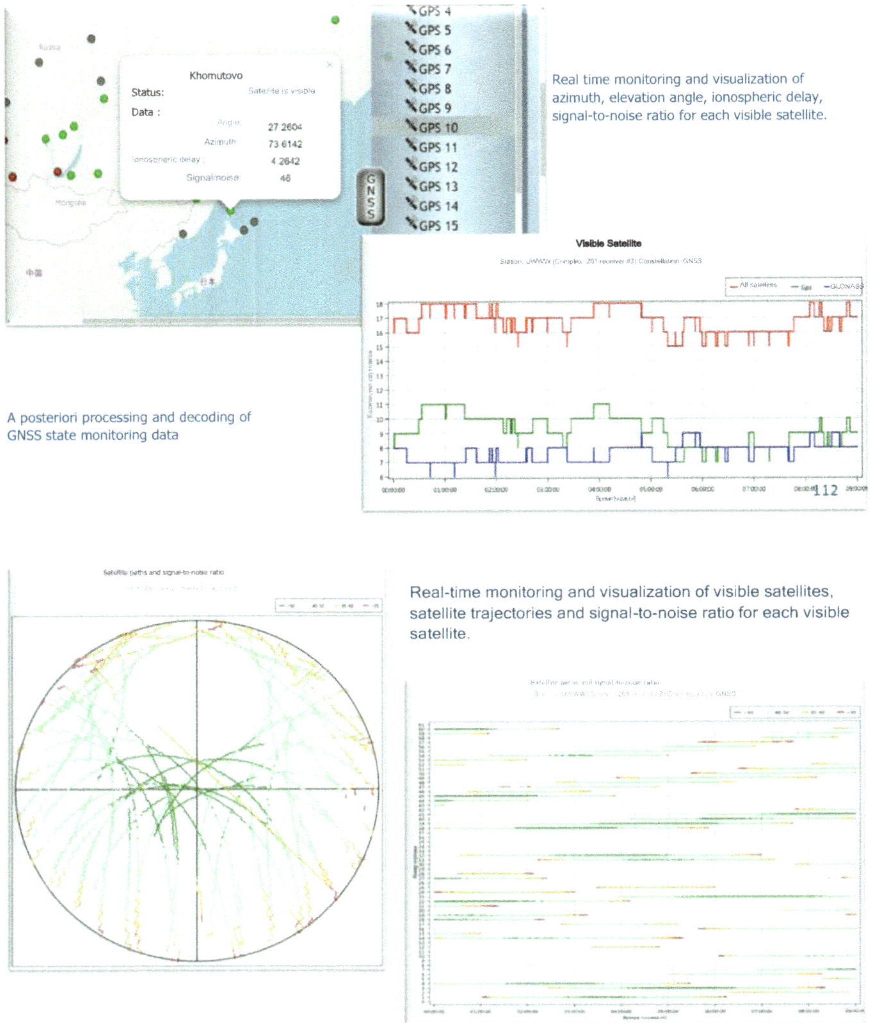

Fig. 3.70 (continued)

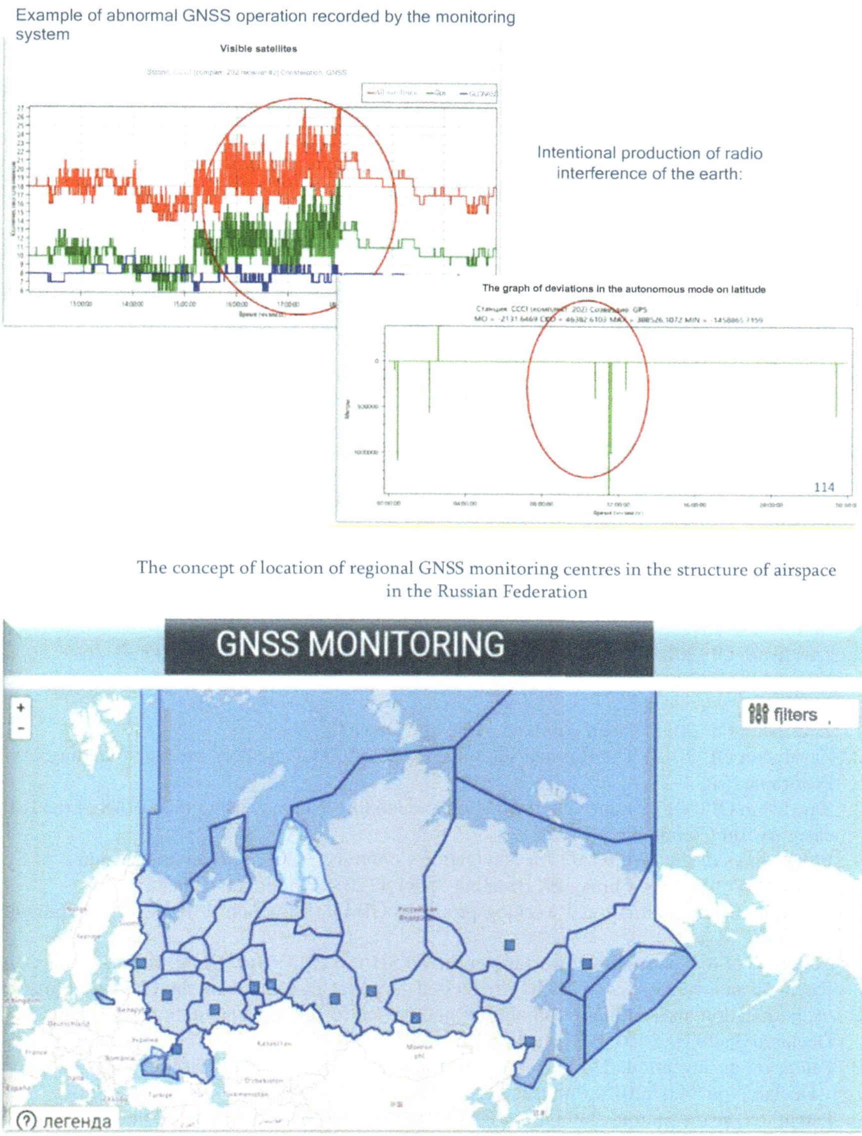

Fig. 3.70 (continued)

References

1. The Certificate of flight tests of LMCS-A- 2000 "On the results of the ground-based airfield and flight tests of satellite navigation and landing equipment prototypes of the II and III categories and the system elements on the IL-76 and IL-103 laboratory airplanes under conditions of actual interference of the airfield" No. 163–11-IX. Gromov Flight Research Institute (2011)
2. GOST 26121–84. Airborne instrumental approach systems for landing (1984) Terms and definitions. — M.: Standards Publishing, p 8
3. Zavalishin OI (2007) High-precision navigation from "RDPF Spektr". Bull Aviation Cosmonautics (2):31
4. Zavalishin OI (2017) Ground and flight inspection complexes developed by LLC RDPF SPEKTR. Zavalishin OI, Zatuchny DA (eds) Inf Commun (2):27–31
5. Zavalishin OI (2003) Monitoring of the quality of signals at the control-correction stations, which provide for the differential mode of operation of users of satellite radio navigation systems. Zavalishin OI, Korchagin VA, Lukoyanov VA, Mironov MA (eds) Radio Eng (1):35–44
6. Zavalishin OI (2002) Local monitoring and correction station LMCS-A-2000. In: Collection of scientific papers based on the materials of the 1st International Radio-Electronic Forum "Applied Radio Electronics. The status and development prospects of the MRF-2002", pp 40–42
7. Zavalishin OI (2009) Local functional extensions to GNSS "RDPF Spektr". Navigation News 4:25–31
8. Zavalishin OI (2007) The results of testing satellite equipment LMCS-A-2000 and GLS in Europe. Zavalishin OI, Lukoyanov VA (eds) Navigation News 4:13–16
9. Zavalishin OI (2018) About two-constellation GBAS. Proc MSTUCA. 3(21):37–46
10. Zavalishin OI (2017) Evaluation of the GBAS ground station operability. Inf Commun (2):22–26
11. Zavalishin OI (2011) Cleared to land. Aerosp Courier (2):71
12. Zavalishin OI (2014) The Russian station LMCS-A-2000 guarantees precision landing. Avia Panorama (5):24–27
13. Zavalishin OI (2017) Improving navigation and landing accuracy using the GBAS of the II/III category. Inf Commun (2):18–21
14. Proceedings of the 12th ICAO Air Navigation Conference (2012)/Montreal, Canada
15. ICAO NSP materials. Flimsy 29, Baseline SARPS (2010)/Montreal, Canada
16. Materials of the international working group on GBAS (Eurocontrol, FAA) (2017)/Entente, USA
17. Materials of the international working group on SBAS (2014)/New Delhi, India
18. Patent on an invention. 2441203 RF. Integrated universal all-weather method for determination of the location and landing of an aircraft and device for its implementation/Zavalishin O.I.–Declared on 29.09.2010; Published on 27.01.2012, Bul. No. 3
19. Patent on an invention. 2542325 RF. The method of location of an aircraft / Zavalishin OI – Declared on 24.07.2013; Published on 20.02.2015, Bul. No. 5
20. Patent on an invention. 2393504 RF. The method of determination of the unacceptable anomalies of the received signals of navigation satellites and the device for its implementation/Zavalishin O.I.–Declared on 23.10.2008; Published on 27.06.2010, Bul. No. 18
21. Patent on an invention. 2541691 RF. The way to improve the integrity of the output signals of on-board satellite navigation receivers / Zavalishin O.I., Lebedev B.V. – Declared on 12.08.2013; Published on 20.02.2015, Bul. No. 5
22. Patent on an invention. 2542326 RF. The way to improve the integrity of the used signals of navigation satellites using a local control and correction station, taking into account the influence of the anomalous ionosphere / Zavalishin O.I. – Declared on 04.10.2013; Published on 20.02.2015, Bul. No. 5

23. Patent on an invention. 2550907 RF. The method of landing an aircraft using ultraviolet receivers and emitters (2 variants) / Zavalishin O.I. – Declared on 04.12.2013; Published on 20.05.2015, Bul. No. 14
24. Patent on an invention. 2488139 RF. The method for reducing the influence of multipath propagation of radio signals of navigation satellites and device for its implementation / Zavalishin O.I. – Declared on 27.05.2011; Published on 20.07.2015, Bul. No. 20
25. Patent on an invention. 2497079 RF. The method of photon location of an air object / Zavalishin O.I. – Declared on 07.06.2012; Published on 27.10.2013, Bul. No. 30
26. Patent on an invention. 2659469 RF. The method for global monitoring of the region livelihood through a single network of local control and correction stations / Zavalishin O. I. – Declared on 13.09.2017; Published on 02.07.2018, Bul. No. 19
27. Patent on an invention. 2653066 RF. The method of determination of the ionospheric storm using ground stationary LMCS / Zavalishin O.I. – Declared on 29.08.2017; Published on 07.05.2018, Bul. No. 13
28. Volume 1. Radio navigation equipment. Annex 10 to the Convention on International Civil Aviation: Aviation telecommunication. International Civil Aviation Organization, 2006. p 616
29. Certificate of state registration of the computer program "Software of the Ground and Flight Support Complex of the GNSS Functional Extension" No. 20136614074; Published on 20.05.2016, [3
30. Federal Aviation Rules "Air Traffic Management in the Russian Federation". Approved by the Order of the Ministry of Foreign Affairs of the Russian Federation from 25.11.2011 No. 293/Garant. Information and legal support [Electronic resource]. https://base.garant.ru/70117238/.
31. Federal Aviation Rules "The procedure for the implementation of radio communications in the airspace of the Russian Federation." Approved by the Order of the Ministry of Foreign Affairs of the Russian Federation from September 26, 2012 No. 362/Consultant Plus [Electronic resource] - https://www.consultant.ru/document/cons_doc_LAW_145093/.
32. Federal Aviation Regulations "Radio technical support of aircraft operations and aviation telecommunications in civil aviation" approved by the Order of the Ministry of Foreign Affairs of the Russian Federation from 20.10.2014 No. 297/Garant. Information and legal support [Electronic resource]. https://base.garant.ru/70812462/.
33. Dou B (2011) Navigation Satellite system. Signal in space interface control document (Test Version). China Satellite Navigation Office
34. Birnbaum ZW, Esary JD, Saunders SC (1961) Multicomponent system and structures and their reliability. Technometrics 3(1)
35. ICAO: International Standards and Recommended Practices. ANNEX 10. vol. 1 (Radio Navigation Aids), 1999–2006
36. Zavalishin OI (2010) The test results of GBAS of the II / III category pf development of LLC "RDPF Spektr". Navigation News 4:24–31
37. Fedosov DV (2003) Precision flight control methods for advanced landing systems: thesis for the degree of candidate of Technical Sciences: 05.22.13. Moscow, p 135

Chapter 4
Improvement of Noise Immunity of Navigation Systems of Aircraft of Civil Aviation on the Basis of Satellite and Inertial Navigation System and Also Ground-Based Systems

4.1 The Algorithm for Detection of "False Satellites" Based on the Integration of Satellite and Inertial Navigation Systems

For the case when the reflected signal does not differ from the true one by more than 8.5–10 dB, the dynamic range of the level changes in the SRNS, which depends on the difference between the conditions in the transmission of navigation signals from SC, which form the working constellation, is approximately the same. It is possible to make the conclusion about the high probability of registration of navigation messages received from "false satellites" [1].

The errors due to false navigation messages generated in a similar way cannot unlike anomalous errors caused by abnormalities in SC operation, but can be automatically determined by using the RAIM function, which provides integrity data in the autonomous mode for the SRNS receiving indicator. It should be noted that the integrity for such a case should be understood as the degree of reliability to the obtained navigation data. This information includes the following components: the message that the error in an aircraft positioning exceeded a certain prestart value, the probability of receiving incorrect information and the notification time for SC operation in an abnormal mode.

Integration (complex processing) of navigation information received from the SRNS and INS is one of the ways to subtract imitational navigation interference (signals from "false satellites"). The idea of this method is to assess the differences in the values of radio navigation parameters (RNP), calculated on the basis of determination of an aircraft location using the INS and determined by the SRNS. Under certain assumptions, this method is a certain analogue of the method of improving the RAIM function used in the SRNS–INS systems obtained by integration [2, 3].

The rates of change of the coordinates of an aircraft, rather than the coordinates of an aircraft themselves, are the main object of observation for differences in the

O. I. Zavalishin et al., *Modern Requirements for Noise Immunity Aircraft Navigation Equipment*, Springer Aerospace Technology,
https://doi.org/10.1007/978-981-16-0073-9_4

values of the radio navigation parameters calculated on the basis of determination of an aircraft location using the INS and determined by the SRNS. Since in the INS the main data is the rate information, this fact is reliable. Using this information, it is possible to determine an aircraft location. It should be noted that the accumulation of errors in latitude and longitude will be quite large due to the low efficiency of the SRNS.

Let us consider the following method. During normal operation in normal mode of all NSC of the SRNS and INS working constellation at the time instant t_k, the vector of differences of rates Δv_k is created, which is measured by the INS $v_k^{ИНС}$and the SRNS receiver v_k^{CHC} at the same time instant t_k as follows

$$\Delta v_k = v_k^{CHC} - v_k^{ИНС}. \tag{4.1}$$

The calculated value Δv_k enters the Kalman filter. Let us introduce the indication:

$$\Phi = \exp(Ft_d), \tag{4.2}$$

the transition matrix, which is one of the components of the equation of an aircraft, written as follows:

$$\Delta v_k = \Phi_{k-1} \Delta v_{k-1} + W_k, \tag{4.3}$$

where W_k is the vector, which is a Gaussian random process with a covariance matrix Q;

$$F = \begin{bmatrix} 0 \ldots 1 \ldots & 1 \\ 0 \ldots \quad 0 & 1 \\ -P_0 \ -P_1 & -P_2 \end{bmatrix}, \tag{4.4}$$

where $P_0 = 1,725612 \cdot 10^{-6}$, $P_1 = 26,18236 \cdot 10^{-4}$, $P_2 = 1$; $t_d = 0,1$ and c is the discretization interval received in the SRNS receiver.

Let us introduce the matrix of partial derivative measurements of pseudo range P K_{KA_i} to SC on the determined coordinates of an aircraft in the geocentric coordinate system $X = [x, y, z, \delta t]$, where δt is the shift of the time scale of an aircraft in relation to the system time [4]:

$$H = \begin{bmatrix} \frac{x - x_{KA_1}}{PR_{KA_1}} & \frac{y - y_{KA_1}}{PR_{KA_1}} & \frac{z - z_{KA_1}}{PR_{KA_1}} & 1 \\ \frac{x - x_{KA_2}}{PR_{KA_2}} & \frac{y - y_{KA_2}}{PR_{KA_2}} & \frac{z - z_{KA_2}}{PR_{KA_2}} & 1 \\ \ldots & \ldots & \ldots & \ldots \\ \frac{x - x_{KA_n}}{PR_{KA_n}} & \frac{y - y_{KA_n}}{PR_{KA_n}} & \frac{z - z_{KA_n}}{PR_{KA_n}} & 1 \end{bmatrix} \tag{4.5}$$

The vector of observations (measurements), which is a Gaussian random process with the covariance matrix R, can be written in the following form:

$$z_k = H_k x_k + n_k, \tag{4.6}$$

$$P_k = M\left[\left(\Delta v_k - \Delta\hat{v}\right)\left(\Delta v_k - \Delta\hat{v}\right)^T\right]^R, \tag{4.7}$$

which is the covariance matrix of estimation errors $\Delta\hat{v}_k$, where M is the averaging symbol, K_k is the gain matrix of the Kalman filter, $P_{k/k-1}$ is the predicted covariance matrix for the predicted estimate $\Delta\hat{v}_{k/k-1}$.

At the output of the filter, the best linear estimate $\Delta\hat{v}_k$ is observed. The presence of the value $\Delta\hat{v}_k$ allows to find the best estimate of the rate $\Delta\hat{v}_k$ by measurements of the two considered navigation sensors according to the expression [4]

$$\hat{V}_k = V_k^{ИНС} + \Delta\hat{V}_k, \tag{4.8}$$

where $\hat{v}_k = \begin{bmatrix} \widehat{V}_k^N \\ \widehat{V}_k^E \end{bmatrix}$, $\Delta\hat{v}_k = \begin{bmatrix} \widehat{V}_k^N \\ \widehat{V}_k^E \end{bmatrix}$, $V_k^{ИНС} = \begin{bmatrix} V_{N_k}^{ИНС} \\ V_{E_k}^{ИНС} \end{bmatrix}$.

The best estimate of coordinates $\hat{x}_k = \left[\hat{\varphi}_k, \hat{\lambda}_k\right]$ can be found by numerical integration:

$$\begin{gathered} \hat{\varphi}_k = \widehat{\varphi_{k-1}} + \sum_{k=1}^{N} \frac{\hat{V}_k^N}{M}(t_k - t_{k-1}),\ \hat{\lambda}_k = \widehat{\lambda_{k-1}} + \sum_{k=1}^{N} \frac{\hat{V}_k^E}{N\cos(\hat{\varphi}_k)}(t_k - t_{k-1}), \\ M = (a+H)\frac{1-e^2}{\left[1 - e^2\sin^2(\hat{\varphi}_{i-1})\right]^{3/2}},\ N = (a+H)\frac{1}{\left[1 - e^2\sin^2(\hat{\varphi}_{i-1})\right]^{1/2}}, \end{gathered} \tag{4.9}$$

where M and N are the radii of curvature of the meridian and the first vertical, respectively, H is the AC flight altitude, a is the major semi-axis of the earth's ellipsoid.

The found estimate $\Delta\hat{v}_k$ goes to the prediction block, which is an open Kalman filter. In the forecast block, this value is extrapolated to the point in time $t_k + \Delta t$, where $\Delta t = nt_d$

$$\Delta v_{k+n} = \Phi\Delta\hat{v}_k + Q. \tag{4.10}$$

The value Δv_{k+n} represents the predicted value of the discrepancies between the measurements of the rate of RMD of the SRNS and the INS. The equivalent discrepancy in coordinates can be found by numerical integration of the resulting vector Δv_{k+n}. The corresponding expressions (4.9) should be written in the form [5]:

$$\Delta\phi_{k+n} = \sum_{i=1}^{n} \frac{\Delta V_{k+i}^{N}}{M} t_d,\ \Delta\lambda_{k+n} = \sum_{i=1}^{n} \frac{\Delta V_{k+i}^{E}}{N\cos\left(\hat{\phi}_k\right)} t_d,\ \Delta x_{k+n}^{ИНС} = \begin{bmatrix} \Delta\phi_{k+n} \\ \Delta\lambda_{k+n} \end{bmatrix} \qquad (4.11)$$

where ΔV_{k+i}^{N} and ΔV_{k+i}^{E} are the rate differences in latitude and longitude.

The equation $\Delta v_{k+n} = \Phi \Delta \hat{v}_k + Q$ makes it possible to predict the difference in the rate registered by the RMD of the INS and SRNS. At the same time, the found transition matrix Φ includes most of the error components of the INS for which it was obtained.

The analysis of the experimental flight data shows that the noise component of the process of rate differences determined by the RMD of the SRNS and INS is a Markov process, having the correlation time of about 1 s and the dispersion $\approx 0{,}0016\ м^2/c^2$for platform INS and $\approx 4 \cdot 10^{-4}\ м^2/c^2$for frameless INS.

Thus, under the assumption about uncorrelatedness of horizontal channels N and E, the equivalent random component of the divergence in coordinates $\delta_{\Delta x}$ for the forecast interval Δt can be determined from the expression

$$\delta_{\Delta x}^2 = \frac{\Delta t}{t_d} \cdot \delta_{\Delta v}^2. \qquad (4.12)$$

When forecasting on the interval $\Delta t = 2$ min and with the sampling interval $t_d = 0,1\ t_d = 0,1$ s, the input noise will be of the order $2\ м^2$.

The INS leaving in relation to the SRNS by coordinates during the prediction time should be determined. As a simplified model of difference in rates $\Delta v(t)$ it makes sense to use a harmonic model:

$$\Delta v(t) = A\cos(\omega_ш t + \phi), \qquad (4.13)$$

where $\omega_ш \approx 0{,}00124\ рад/c$is the Schuler frequency, A and φ are the amplitude and the initial phase of the fluctuations of INS leaving in relation to the SRNS.

In the worst case, when during the extrapolation interval $cos(\omega_ш t + \varphi) = 1$, the value Δv is maximum. The corresponding value Δx cc can be written as

$$\Delta x(t_0, t_1) = \int_{t_0}^{t_1} A dt = A(t_1 - t_0) = A \cdot \Delta t. \qquad (4.14)$$

For the case of platform-type INS, the generalization over many flights shows that the amplitude of the oscillatory process is within $1 \div 1{,}3\ м/c$. This means that at the interval $\Delta t = 2$ min at the time of sampling $t_d = 0,1c$ the discrepancy in coordinates, determined by means of the RMD of the SRNS and INS, in the worst case may be about 160 m.

It is possible to conclude that at maximum values Δv the contribution of the noise component is less noticeable; however, when crossing Δv zero, its contribution is comparable to the influence of the oscillatory process.

Thus, at the moment of time t_k the predicted value of INS leaving by coordinates in relation to the RMD of the SRNS Δx_{k+n} is generated valid for the extrapolation interval Δt.

Starting from the moment t_k during the time interval Δt c with the frequency of updating of the navigation information, the current INS coordinates are recalculated taking into account the observed discrepancies in the SRNS and INS coordinates $\Delta x_{k+i}^{СРНС-ИНС} = \left(\varphi_{k+i}^{СРНС=ИНС}, \lambda_{k+i}^{СРНС-ИНС}\right)^T$ and the predicted value $\Delta x_{k+n} = (\Delta\varphi_{k+n}, \Delta\lambda_{k+n})^T$ $\Delta x_{k+n} = (\Delta\varphi_{k+n}, \Delta\lambda_{k+n})^T$ into equivalent pseudo ranges to all SC of the working constellation $R_{k+i}^{ИНС}$:

$$R_{k+i}^{ИНС} = \sqrt{(X_{k+i}^{КА} - X_{k+i}^{ИНС})^2 + (Y_{k+i}^{КА} - Y_{k+i}^{ИНС})^2 + (Z_{k+i}^{КА} - Z_{k+i}^{ИНС})^2} \tag{4.15}$$

In Eq. (4.15) $X_{k+i}^{ИНС} = \left(x_{k+i}^{ИНС}, y_{k+i}^{ИНС}, z_{k+i}^{ИНС}\right)^T$ is the vector of INS coordinates in the geocentric coordinate system:

$$\begin{aligned}
x_{k+i}^{ИНС} &= \left(N + h_{k+i}^{БАРО}\right)\cos\left(\varphi_{k+i}^{ИНС} - \varphi_{k+i}^{СРНС-ИНС} + \Delta\varphi_{k+n}\right)\cos\left(\lambda_{k+i}^{ИНС} - \lambda_{k+i}^{СРНС-ИНС} + \Delta\lambda_{k+n}\right),\\
y_{k+i}^{ИНС} &= \left(N + h_{k+i}^{БАРО}\right)\cos\left(\varphi_{k+i}^{ИНС} - \varphi_{k+i}^{СРНС-ИНС} + \Delta\varphi_{k+n}\right)\sin\left(\lambda_{k+i}^{ИНС} - \lambda_{k+i}^{СРНС-ИНС} + \Delta\lambda_{k+n}\right),\\
z_{k+i}^{ИНС} &= \left((1-e)^2 N + h_{k+i}^{БАРО}\right)\sin\left(\varphi_{k+i}^{ИНС} - \varphi_{k+i}^{СРНС-ИНС} + \Delta\varphi_{k+n}\right),\\
N &= \frac{a}{\sqrt{1 - e^2 \sin^2(\hat{\varphi})}},
\end{aligned} \tag{4.16}$$

where $h_{k+i}^{БАРО}$ is the barometric height, according to the SRNS RMD at the point in time t_k. We note that at the moment of time t_k it is assumed that all SC of a working constellation, as well as INS are operating normally.

The predicted pseudo range vector based on information from the INS $R_{k+i}^{ИНС}$, as well as the vector of pseudo ranges measured by the SRNS RMD $PR_{k+i}^{СРНС}$ at each time point $t_i = i \cdot t_d$ during the current forecast interval $\Delta t = nt_d$ go to the pseudo range difference analysis block.

In this block, the vector is formed, which is the difference of vectors $R_{k+i}^{ИНС}$ and $PR_{k+i}^{СРНС}$

$$\Delta R_{k+i} = R_{k+i}^{ИНС} - PR_{k+i}^{СРНС},$$

$$\Delta R_{k+i} = \begin{bmatrix} \Delta r(1)k + i \\ \Delta r(2)^{k+i} \\ ----- \\ \Delta r(j)k + i \\ ------ \\ \Delta r(m)k + i \end{bmatrix}, \tag{4.17}$$

where m is the number of SCs that are in tracking mode in the measuring channels of the SRNI RMD.

The pseudo ranges $\Delta r(j)^{k+i}$ of the vector of differences ΔR_{k+i}, obtained by processing the information of the navigational instruments under consideration can be chosen as diagnostic signs for detecting "false satellites" [6].

Let us suppose the situation where at the point in time t_k, within the interval Δt, that is at $t_k \le t_e \le t_k + \Delta t$, at the next update of information, a "false satellite" is captured as the jth SC of the working constellation. This leads to the appearance of an abnormally large component $|\Delta r_e(j)_{k+i}|$ of the vector ΔR_{k+i}, corresponding to the jth measuring channel. If the jth component exceeds the threshold corresponding to the permissible level of coordinate determination errors in the horizontal plane [horizontal alert level (HAL)], specified for different phases of the flight of the aircraft by the standards

$$|\Delta r_e(j)_{k+i}| \ge \Delta r_{MAX},$$

then the decision is made on the presence of an anomalous measurement in the jth measuring channel of the SRNS RMD and the jth SC is excluded from the solution of the navigation task.

Let us find the expression for the threshold Δr_{MAX}. This value, on the one hand, is a function of the HAL parameter, that is $\Delta r_{MAX} = f(HAL)$; on the other hand, it is possible to write

$$\Delta r_{MAX} = H_1 \begin{bmatrix} \Delta N \\ \Delta E \\ \Delta H \end{bmatrix} = H_1(j, 1) \cdot \Delta N + H_1(j, 2) \cdot \Delta E, \tag{4.18}$$

where $H_1(j, 1)$ and $H_1(j, 2)$ are the elements of the matrix (4.18) at the intersection of jth line with the 1st and 2nd columns, respectively, and ΔN and ΔE are the maximum allowable error levels for determination of latitude and longitude. In this case, the maximum allowable error in determination of the horizontal coordinates can be determined from the expression

$$\sqrt{\Delta N^2 + \Delta E^2} = HAL. \tag{4.19}$$

From the expression (4.19) we write the expression for the component ΔE

$$\Delta E = \sqrt{HAL^2 - \Delta N^2}. \tag{4.20}$$

Making the substitution of the expression (4.20) into (4.18) and differentiating the resulting expression on ΔN, we obtain

$$\Delta r_{M\alpha X} = H_1(j, 1) - \frac{H_1(j, 2)}{\sqrt{HAL^2 - \Delta N^2}}. \tag{4.21}$$

Equating the Eq. (4.21) to zero and solving it in relation to ΔN, we get the following expression for ΔN

$$\Delta N = HAL \frac{H_1(j, 1)}{\sqrt{H_1(j, 1)^2 + H_1(j, 2)^2}}. \tag{4.22}$$

After substituting (4.22) into (4.20) and doing simple transformations, we obtain as a result the expression for ΔE in the form

$$\Delta E = HAL \frac{H_1(j, 2)}{\sqrt{H_1(j, 1)^2 + H_1(j, 2)^2}}. \tag{4.23}$$

Thus, substituting the obtained expressions for ΔN and ΔE in (4.18), we obtain the desired expression for Δr_{MAX}:

$$\Delta r_{MAX} = HAL \cdot \max\left(\sqrt{H_1(j, 1)^2 + H_1(j, 2)^2}\right), \tag{4.24}$$

where $j = 1, 2, \ldots, m$.

Diagnosing is checking of the condition:

$$\Delta R^J_{k+i} \geq \Delta r_{MAX}. \tag{4.25}$$

The obtained ratio for the threshold in the form (4.24) allows to answer the question of the presence of an anomalous measurement, which leads to errors comparable to the maximum errors in determination of the HAL horizontal coordinates for the current phase of an aircraft flight. An anomalous pseudo range measurement, the cause of which does not lead to errors exceeding HAL, in diagnosing, which consists in comparing each component of the vector of differences ΔR_{k+i} with the value (4.24), will be qualified as normal.

To detect small jumps of pseudo ranges, the predicted value of the divergence of the horizontal coordinates of the SRNS and INS RMD $\Delta x^{ИНС}_{k+n}$ should be used as the maximum error. In this case, the expression (4.19) can be written as

$$\sqrt{\Delta N^2 + \Delta E^2} = \Delta x^{ИНС}_{k+n}. \tag{4.26}$$

Taking into account the above, we write the expression to detect small errors in pseudo ranges:

$$\Delta r_{П} = \Delta x_{k+n}^{ИНС} \cdot \max\left(\sqrt{H_1(j,1)^2 + H_1(j,2)^2}\right). \tag{4.27}$$

Let us consider the statistical properties of the vector of differences ΔR_{k+i}. This vector is the result of the action of noises of measurements of pseudo ranges by the SRNS RMD with the covariance matrix $\delta_0^2 I$, where I is the single matrix of the dimension $m \times m$, and the noise component of the change in the difference of the speeds of the SRNS and INS RMD. Thus, the pseudo range difference vector, measured by the SRNS RMD and calculated on the basis of the INS navigation data, is a normal random vector with the expectation determined by the value of the predicted coordinate determination discrepancy $\Delta x_{k+n}^{ИНС}$. Taking into account the above, we will write down the statistical threshold for the detection of anomalous measurements in the vector ΔR_{k+i} with the set probability of false alarm $P_{л.m.}$:

$$\Delta r_{П} = \left(\Delta x_{k+n}^{ИНС} + x_{П} \cdot \delta_{\Delta x}\right) \cdot \max_{j}\left(\sqrt{H_1(j,1)^2 + H_1(j,2)^2}\right) + x_{П} \cdot \delta_0, \tag{4.28}$$

where $x_{П}$ and $P_{лm}$ are connected with the correlation

$$P_{л.m} = 1 - F(x_n), \tag{4.29}$$

$F(x_n) = \int_{-x}^{x_n} \frac{1}{\sqrt{2\pi}} e^{-\frac{x^2}{2}} dx$ is the integral normal distribution function (tabulated Laplace function), $\delta_{\Delta x}$ is the root mean square value of the random component of the discrepancy in the coordinates during the prediction interval, defined by the expression (4.12).

Diagnosing in this case consists in checking of the following condition:

$$\Delta R_{k+i}^{j} \geq \Delta r_{П}. \tag{4.30}$$

Thus, if the condition (4.25) is not satisfied for any of the components of the vector of differences ΔR_{k+i}, this means that there are no anomalous measurements or their level is below the response threshold. Further diagnosis in this case can be carried out by checking the condition (4.30).

If the matrix $H1$ is defined in the geocentric coordinate system xyz, it must be transformed into the coordinate system NEH, connected with an aircraft. To do this, it is necessary to perform the following matrix transformations:

$$H_{NEH} = H_{xyz} \cdot T^T, \tag{4.31}$$

where $T = \begin{bmatrix} T_1 & 0^T \\ 0 & I \end{bmatrix}$.

The matrix T_1 is an orthogonal matrix that transforms the vector X, defined in the geocentric coordinate system into a local coordinate system whose center is at the point with the coordinates $X = [x, y, z]^T$:

$$T_1 = \begin{bmatrix} -\sin(\varphi)\cos(\lambda) & -\sin(\varphi)\sin(\lambda) & \cos(\lambda) \\ -\sin(\lambda) & \cos(\lambda) & 0 \\ \cos(\varphi)\cos(\lambda) & \cos(\varphi)\sin(\lambda) & \sin(\varphi) \end{bmatrix}. \tag{4.32}$$

I is a single matrix with the dimension (2×2) in case the SRNS RMD solves the navigation task using the GLONASS and GPS SRNS with an estimate of two time parameters, or (1×1) when working with only SRNS with an estimate of one time parameter. The matrix 0 is a zero matrix of the dimension (3×1) if one time parameter is estimated, or (3×2) if two time parameters are estimated [7].

Thus, if in the difference vector (4.17), the $\Delta R_{k+i}\,j$th component exceeds the value $HAL \cdot \max_j\left(\sqrt{H_1(j,1)^2 + H_1(j,2)^2}\right)$, then the decision is made about the presence of an anomalous measurement in the jth SRNS RMD channel.

4.2 Improving the Noise Immunity of Aircraft Navigation Systems in Subpolar Latitudes Based on the Use of Aircraft Positioning Devices on Satellite Radio Navigation Systems and Ground-Based Radio Navigation Systems

The region of the Extreme North is the zone of special attention for many countries. One of the reasons for this is that the shortest air routes connecting various regions of the globe pass through it.

It should be noted that, based on the available information, it can be concluded that the shortest route lengths that connect Southeast Asia with North America through the Arctic Ocean, the Far East and Eastern Siberia do not meet modern requirements for AC flight safety.

One of the reasons for this is that modern means of navigational support for aircraft flights, which are currently used in aviation operating in polar latitudes, are not able to provide regulatory requirements in the event of the influence of problems connected with these areas, arising due to different layers of the atmosphere, and also because of the different underlying surface.

Another problem is the large territory of the Extreme North. As a result, it is difficult to place the necessary number of radar positions on it. Therefore, in the polar latitudes at the present time it is practically not possible to carry out continuous radar surveillance of the AC movement.

Disadvantages of determination of the location of aircraft based on the SRNS in polar latitudes are as follows:

(1) Low accuracy characteristics due to poor geometrical factor of working constellations caused by the low location of the NSC above the horizon;
(2) The influence of the underlying surface due to the fact that the NSC transmits navigation signal in the decimeter wave band;
(3) Low NSC signal power and the possibility of its suppression by a terrorist group.

Based on the above, it is necessary to solve the problem of the SRNS common use with other navigation systems for the polar latitudes, or even alternative use of these systems.

At present, navigation support for aircraft in high latitudes is mainly carried out using inertial and course systems. For these areas, automatic radio compasses which are located on the aircraft and operate in the VHF waveband are also widely used. This equipment has the ability to work in the range of meter and decimeter waves. This makes it possible to ensure a meeting of helicopters and airplanes in the airspace, to exchange the navigational information between the aircraft in the event of a flight of an aircraft belonging to the group, as well as when searching for the source of the aircraft signal requesting help. The analysis of statistical data shows that the functioning of these means is impaired significantly due to the phenomenon of multi-path, i.e., re-reflections of signals from the troposphere, the ionosphere and the underlying surface of the earth.

In the polar latitudes, it is also difficult to use short-range navigation tools, since it is difficult to place power stations with large capacity and their service personnel on ice floes and small islands.

In order to provide navigation requirements for aircraft flights in polar latitudes, it is recommended to use the communication in all wavelength ranges from VHF to KMV waves. In addition, the VHF band is used to provide near, direct line-of-sight, communications. For the purpose of providing communication over long distances it is recommended to use a DMW-waveband. To ensure the functioning of communications in the near zone with aircraft having radio stations operating in KMV and GKMV-radio wave bands, KMV and GKMV ranges are used.

The analysis of the results of using of VHF-communication shows that in the region of the Extreme North there is no complete overlap of VHF-communication. It should be noted that at certain intervals of the route there are significant discontinuities in the connected field. Thus, the requirements for the continuous operation of UHF communications in this case are not met.

It should be noted that for the polar latitudes the DMW communication is unstable, due to the impact of large numbers of protons and electrons on the ionospheric layers that reflect radio waves during periods of ionospheric and magnetic storms.

Due to the absence of high-power transmitters on the earth's surface in the polar latitudes and participation in the process of navigation support for the aircraft, it is impossible to use full-wave KMV and GKMV. In addition, in the present time, it is required to upgrade the polar medium-wave and long-wave drive radio stations operating in polar latitudes.

For the zone of the polar troposphere, it is necessary to take into account the strong influence of inhomogeneities that appear in the case of large negative temperatures in winter.

It should be noted that the inhomogeneities of the underlying surface have a great influence on the navigation support of aircraft operations in the tundra and in the ice of the Arctic Ocean. In relation to different seasons, the range of operation of navigation support systems for aircraft operations depends on the type of underlying ground surface.

To improve the reliability and accuracy of navigation definitions in polar latitudes, long-wave RNS are also used.

In the domestic analog of the long-wave RNS "Loran-S" the signal is a periodic sequence of narrow radio pulses with the leading edge duration $\tau_{\phi} = 65\,мкс$. The RMSE of determination of the location does not exceed 300–500 m (above the sea). Above the land the accuracy decreases. The frequency band of the pulse signal of the pulse-phase RNS "Laurent-C" is approximately 7500 Hz at the level of 0.7.

The signal of "Loran-S" is a radio pulse, whose envelope is described by the expression

$$s(t) = \left(\frac{t}{t_u}\right)^2 \exp\left(2 - 2\frac{t}{t_u}\right), \tag{4.33}$$

where t_u is the pulse front duration, $t_u = 65\,мкс$. Carrier frequency of radio pulses emitted by reference stations $f_0 = 100\,кГц$.

The ground stations of the Loran-S system emit signals at a single carrier frequency. In order to avoid mutual interference among the stations of the chain, the temporary separation of signals is used, with each station emitting a pack of 8 pulses with intervals of 1 ms between them.

In the pulse-phase RNS "Laurent-S" for the resolution of the ambiguity of measuring of the time of arrival of the signal (phase) at the carrier frequency, the time of arrival of the signal is measured by the envelope of the radio pulse.

The following methods are most known for determination of coordinates by the phase method:

(1) Range-measurement method with highly stable generators;
(2) Range-measurement method with re-emission (RNS are almost never used in the long wave);
(3) Difference between range-measurement methods.

In the range of long radio waves, the electromagnetic energy of a signal reaches its receiving point in two ways: propagating along the earth's surface (surface wave) and being repeatedly reflected from the ionized layer of the ionosphere (spatial wave).

To ensure the required accuracy of phase measurements and the stable elimination of the ambiguity of phase counting, it is necessary to ensure reliable selection of signals of spatial and surface waves. The main parameters that must be taken into account are the ratio of the amplitudes of these signals and the delay time of the spatial wave.

The surface wave signal characteristics, which determine the accuracy of phase measurements, are distinguished by good stability and are mainly subject to seasonal changes connected with an increase or decrease in the rate at which radio waves propagate. These changes should be taken into account as amendments calculated by certain algorithms.

The most unstable are the characteristics of the spatial wave signal. They are subject to changes during the day and are determined mainly by the state of the ionosphere layer reflecting radio waves.

Errors caused by spatial radio waves can be significantly reduced by using the fact that the spatial wave lags behind the surface wave by 45–80 μs at distances up to 500 km and 30–45 μs at distances up to about 2500 km. This allows, using narrow radio pulses, to cut off that part of the signal where the spatial wave acts, and to perform phase measurements on the part of the surface signal not affected by the spatial wave. The formation of such a pulse requires a significant expansion of the frequency spectrum to be emitted and received.

The use of this method of separation of the spatial and surface waves does not allow the use of a signal maximum, which is of fundamental importance when operating in the presence of strong noise or solving hidden problems.

For modern phase RNS, the separation of spatial and surface waves is carried out on the basis of regression processing methods (approximation of functions using the least squares method (LSM)).

The LSM allows to perform the measurement operation of the desired parameter on the envelope of the correlation function of the total signal. The existing algorithms for resolving the multiplicity use the maximum of the envelope of the correlation function of the surface signal as the characteristic point of the envelope. This increases the output signal-to-noise ratio compared to the traditional method of resolving the ambiguity (the method of time selection), when the characteristic point of the envelope is selected at the leading edge of the correlation function (in order to reduce the influence of the spatial signal). Therefore, the probability of correct resolution of ambiguity is increased.

Let us indicate by A_0 and A_1, respectively, the values of the maxima of the correlation function of the surface and spatial signals. Then, the estimates for these parameters are taken to minimize the value

$$U = \sum_{i=1}^{m} [\psi_{\Sigma i} - (A_0 S_i + A_1 P_i)]^2, \tag{4.34}$$

where m is the number of discrete points taken in the area of the maximum of the correlation function of the total signal with the interval T_0; $\psi_{\Sigma i}$ is the value of one of the quadrature components of the correlation function of the received signal at the discrete points; S_i and P_i are the form factors of the envelope model of the correlation function of the surface and spatial signals at discrete points. The need to use for resolving the ambiguity several points on the envelope of the correlation function is determined by the fact that it allows to average the influence of the side lobes of the correlation function of the spatial signal in the area of the correlation function of the surface signal.

To determine the unknown values A_0 and c A_1, it is necessary to solve the system of linear equations:

$$\begin{cases} \frac{dU}{dA_0} = 0 \\ \frac{dU}{dA_1} = 0 \end{cases} \tag{4.35}$$

Substituting the solution of this system into the expression (4.34) and, having made the transformations, we get:

$$U = \sum_{i=1}^{i=m} (\psi_{\Sigma i})^2 - \frac{A^2C + B^2D - 2ABF}{CD - F^2}, \tag{4.36}$$

where

$$A = \sum_{i=1}^{m} \psi_{\Sigma i} S_i; \quad B = \sum_{i=1}^{m} \psi_{\Sigma i} P_i; \quad C = \sum_{i=1}^{m} P_i^2; \quad D = \sum_{i=1}^{m} S_i^2; \quad F = \sum_{i=1}^{m} S_i P_i. \tag{4.37}$$

The search for the maximum of the function

$$W_c = \frac{A^2C + B^2D - 2ABF}{CD - F^2} \tag{4.38}$$

corresponds to the search for the minimum on the ensemble of possible models. At the same time, each model corresponds to a certain set of coefficients S_i and P_i, depending on the position of the maxima of the envelopes of the correlation function of the surface τ_0^0 and spatial τ_1^0 signals in relation to the reading point, as well as on the ratio of their levels. The similar calculations are made on the other quadrature component of the correlation function of the received signal to determine the component W_s. The task of the LSM algorithm is to find the maximum of the function [8]

$$W = W_c + W_s, \tag{4.39}$$

and, therefore, the determination of the time position of the maxima of the envelopes of the surface and spatial signals.

As a result of the implementation of the LSM algorithm, the parameter τ_0^0 and the delay of the spatial signal in relation to the surface one $\Delta\tau_1^0$ ($\Delta\tau_1^0 = \tau_1^0 - \tau_0^0$) are determined.

The rather long analysis time connected with the need to enumerate the time delay of the spatial signal for the entire range of possible coefficients of the shape of the direct and reflected signals is a significant disadvantage of the LSM algorithm. In real systems, about 120 thousand computational operations (additions and multiplications) are required for the realization of the LSM by one reflection, taking into account the averaging over several processing cycles.

In this regard, it is proposed to separate the signals of surface and spatial waves, as well as a preliminary assessment of the time position of the surface signal based on cepstral analysis and thereby to reduce the range of variable delays in the LSM algorithm. At the same time, the computational complexity of the algorithm is reduced to 5–8 thousand computational operations. In addition, the cepstral processing may be carried out on one of the quadrature components of the received signal.

The basis of the proposed algorithm for cepstral processing is the definition of the cepstrum of the power of a discrete signal:

$$S^{pc}(nT) = \frac{1}{2\pi j}\oint \ln|S(z)|^2 z^{n-1}dz, \tag{4.40}$$

where $S(z)_z$ is the conversion of the data sequence $s(nT)$.

After transformation (4.40) of the signal $s(nT)$, which is a convolution of two functions of time, impulses that do not overlap on the axis of the cepstral time are formed. When using a power cepstrum, the phase spectrum is not taken into account, which greatly simplifies the processing of signals in the phase RNS receiving indicator, since there are no difficulties connected with the deployment of the phase and the elimination of the linear phase component.

In the case of a single reflection and in the absence of distortions of the surface and spatial waves, the signal at the input of the cepstral processing device can be represented as

$$s(nT) = s_0(nT) + \gamma_1 s_0(nT - n_1T), \tag{4.41}$$

where $s_0(nT)$ is the sampled surface wave signal presented as a sequence of samples with increments T; $\gamma_1 s_0(nT - n_1T)$, is the signal of the spatial wave $s_1(nT)$, presented as the delayed for $t_1 = n_1T$ copy of the surface signal.

The cepstral processing algorithm is based on the compensation of the component characterizing the surface wave signal, which makes it possible to obtain directly a cepstrum containing information about the delay of the spatial signal.

Here are the factors that should be considered in the complex when choosing a signal for the long-wave (LW) phase RNS. Assuming that the frequency band allocated for signals is the same, these factors are as follows [6, 7]:

(1) Peak power of the emitted signal;
(2) The necessary frequency band of the transmitter for emitting a signal or its element;
(3) Minimum duration of the signal or its element;
(4) Maximum signal duration.

The signal with discrete frequency modulation (DFM) is characterized by the fact that the required spectral width is formed by a series of shifted spectral bands that most fully meet the requirements of the efficiency of formation, emission and transmission along with the propagation path of radio waves in relation to the features of the LW RNS. DFM has a low peak emission power. The frequency band of the transmitting device at the time of emission of one element of the DFM signal may be many times smaller than for the emission of a multi-frequency (MF) or pulsed signal.

The envelope of the signal function at the output of the synchronized receiver is characterized by the presence of large side lobes on the range axis. So the first and largest of them is only by 13.2 dB below the peak value of the main pulse of the output signal of the synchronized filter. The presence of side lobes on a range significantly impairs the characteristics of detection and measurement of range.

Let us estimate the energy efficiency of the DFM signal with the weighting of its elements by duration and intra-element frequency modulation with synchronized filtering of the elements.

Let the signal elements at the input of synchronized elemental filters have a single amplitude ($a_i = 1$). The duration of the element at the filter inlet, its amplitude at the filter output, the noise at the output of the element synchronized filter are determined by the expressions [4, 9]:

$$T_i = \lambda_i T_0, \tag{4.42}$$

where T_i is the duration of the ith element, λ_i is the weighting factor determined by the selected weight function, T_0 is the duration of the element at the output of the filter synchronized with it.

$$A_i = \sqrt{\frac{T_i}{T_0}} = \sqrt{\lambda_i} \tag{4.43}$$

$$P_{ш} = \frac{N_0}{T_0}, \tag{4.44}$$

where N_0 is the spectral density of white Gaussian noise.

The synchronized filtration operation is linear, so the voltage at the output of the element synchronized filter at the moment the element ends is an additive mixture of signal and noise

$$y_i = A_i + n_i. \tag{4.45}$$

As a result of the summation of the output signals of element synchronized filters with the weights $\mu_i = \sqrt{\lambda_i}$ cthe voltage after the summation unit will be

$$y_\mu = \sum_{i=-m}^{m} y_i \cdot \mu_i, \tag{4.46}$$

where

$$m = \frac{N-1}{2},$$

N is the number of elements of the signal.

In this case the signal component is equal to

$$A_\mu = \sum_{i=-m}^{m} A_i \cdot \mu_i = \sum_{i=-m}^{m} \lambda_i. \tag{4.47}$$

The dispersion of noise after summation, taking into account the uncorrelated noise at the outputs of the elemental filters is

$$\delta^2_{\mu\Sigma} = \delta^2_{ш} \sum_{i=-m}^{m} \lambda_i \tag{4.48}$$

The signal-to-noise ratio after summation is

$$q^2_{I\Sigma} = q_0^2 \sum_{i=-m}^{m} \lambda_i, \tag{4.49}$$

where q_0^2 is the signal-to-noise ratio for the element at which $T_i = T_0$.

We use the expression (4.49) to estimate the signal-to-noise ratio at the output of the synchronized filter of the DFM signal with the duration elements $T_i = T_0$. As the filter is synchronized then $\lambda_i = 1$.

Accordingly, from (4.49) we get

$$q^2_{II\Sigma} = q_0^2(2m+1) = q_0^2 N. \tag{4.50}$$

Let us compare (4.49) and (4.50) at equal signal energies. The elements in both cases have the same single amplitude, i.e., $a_i = 1$.

The DFM signal energy with the elements duration T_0

$$E_{II} = N\frac{T_0}{2}, \tag{4.51}$$

and the energy of the DFM signal with the modulation of elements on the duration

$$E_I = \frac{T_0}{2} \sum_{i=-m}^{m} \lambda_i. \tag{4.52}$$

Equating the energy of the signals, from (4.49) and (4.50) we get:

$$K^2 = 1, \tag{4.53}$$

where K^2 reflects the value of the energy loss caused by the weight processing of the DFM signal, as compared with the synchronized reception of the DFM source signal with the elements of equal amplitudes and durations, i.e., it characterizes the loss of desynchronization.

It follows from the correlation (4.53) that the proposed method of weight processing with modulation of the elements of the DFM signal by duration and intra-element modulation in frequency (or phase) with subsequent synchronized filtering of signal elements in the receiver and summation with weight, providing the necessary low level of side lobes of the envelope of the correlation function of the DFM signal does not have energy losses in the desynchronization, i.e., in the proposed method, synchronized filtering of the DFM signal as a whole is provided, which is not found in any of the known methods of weight processing. In this sense, it is optimal [10].

It should be noted that "Loran-C" has a number of significant disadvantages.

First, the average signal power is by a long way less than the peak one by 100–140 times. To ensure the required signal size in relation to the level of atmospheric noise at the borders of the working area, the emitted power must reach hundreds and thousands of kilowatts, while the average power remains at the level of several kilowatts.

The most difficult technical problem is emission in the long wavelength range in relation to a broadband signal with a short rise time of the leading edge while simultaneously ensuring high efficiency of the antenna system. The solution to this problem is achieved by constructing antenna structures having a central tower with a height of up to 400 m. Grounding is a network of radially located wires 300 m long.

It is obvious that the construction of pulsed-phase RNS stations is a big technical and organizational problem and requires significant investments and operating costs. Difficulties and costs increase many times if the RNS is to be deployed in remote and sparsely populated areas, such as the Extreme North.

The systematic deployment of the RNS ground station chains Mars 75 and their integration with the SRNS and Loran-C RNS will improve the accuracy and safety of navigation definitions in the Extreme North region. The field of this PHC can be effectively used by all types of aircraft.

The deployment of the RNS stations Mars-75 and their operation have confirmed the high mobility and operational reliability, determined by the simplicity of the transmitting devices, as well as by the low energy level of the emitted signal.

The RNS Mars-75 is notable for high noise immunity, reliable resolution of ambiguity, accuracy in determination of coordinates, and practical independence of accuracy from the time of day and year.

Based on the above, the RNS Mars-75 can be used for decisions connected with an aircraft positioning.

RNS Mars-75 provides the accuracy of determination of the location of 60–300 m, depending on the geometric factor. In differential mode, it provides aircraft positioning accuracy not worse than the SRNS in standard mode. The use of Mars-75 is quite possible for an AC route navigation and landing approach.

4.3 The Approach to Determination of the Reliability of Navigation Information Transmitted from a Helicopter Under Various Interference Conditions

Currently, helicopter technology is being widely implemented in various sectors of the national economy, including civil aviation. Helicopters are used for carrying out of various works on the coast, for servicing oil rigs and for delivering people and goods to remote areas, such as taiga, mountainous areas, marshland, etc. Besides helicopters can be used as the only type of equipment for transporting people from remote areas to the nearest airports that communicate with the central regions of the country. Intensive use of helicopter technology leads to the need to solve the problem of ensuring its noise immunity to various kinds of interference, affecting the trouble-free operation of navigation equipment aboard aircraft.

Today, when ensuring the flights of helicopters, the rules of visual flight are used. At the same time, it is necessary to take into account the fact that aircraft of this type have a strong dependence on meteorological conditions. The presence of any deviations from meteorological conditions, which are considered to be acceptable, means an automatic termination of helicopter flights. At present, it is completely unclear how to fly in the fog, and especially between the helipads [11].

There is the suggestion: to fly in the landing zone on GLS using the local control and correction station, and when helicopters are flying along the route, use the enhanced voice exchange. However, this approach has significant disadvantages connected with the irrational use of the frequency resource and the absence of reliability and accuracy of the received information.

In addition, it is difficult to use satellite navigation receivers on helicopters for the following reason. Today, there are two main airlines for the use of helicopters: UTair and Gazpromavia. The specificity of the use of helicopter technology in UTair is the implementation of the principle of globality. However, the helicopters used in this airline carry out navigation definitions without the use of satellite radio navigation systems (SRNS). The main navigation tool on these aircraft is Joctick. Gazpromavia;

in contrast, uses modernized GLONASS and GPS navigation and landing helicopters, but only in its department on its oil drilling rigs.

Thus, to ensure the safe flight of a helicopter along the route in today's conditions, special importance should be given to the reliability of information transmitted during voice exchanges. This circumstance, in addition to natural external factors, can be connected with the unauthorized penetration into the radio exchange of intruders, as well as with the fact that the pilot is in a state of strong emotional stress.

When analyzing the reliability of information transmitted by voice message from the helicopter, it is necessary to proceed from the following provisions [12]:

(1) The sound wave is a carrier of a significant amount of information about the intellectual, physical and psychological state of the object at the current moment;
(2) Parameters of a sound wave can be constantly changed by an object or, in other words, the helicopter pilot has the ability to change his voice constantly.

We list the necessary estimates that need to be distinguished on the basis of a sound wave in order to conclude how reliable this information is:

SPT is the assessment, expressing the presence of relatively high frequencies in the analyzed speech.

SPJ is the assessment, expressing the presence of relatively low frequencies in the analyzed speech.

AVJ is the mathematical assessment, expressing the average frequency among the low frequencies characterized by SPJ. The existing technology uses this technology to study the degree of thought embedded in the speaker's words.

JQ is the mathematical assessment expressing the standard deviation among the lower frequencies in the speech being studied.

MainF is the assessment of the main frequency in the speaker's voice (as a percentage of the marginal contribution of emotions to the wave) and the exact spectrum of analysis of 16 frequencies, different for men and women.

FX is the number of additional frequencies in the analyzed frequency window, which have relatively high assessments compared to MainF.

FFLIC is the number of direction changes in the analyzed frequency window that have a contribution that exceeds the minimum of the MainF set.

SFLC is the number of changes in the sum of the SPT and SPJ assessments of the original sound wave.

ANT is the total percentage of the two highest frequencies in the frequency window being analyzed.

The process of determination of the unreliability of information received by voice message should be divided into several stages [13]:

1. Determination of the difference between this assessment and the main assessment for each individual characteristic.
2. Processing of total assessments for the marginal degree of difference according to the instructions, indicating that the information is unreliable and that the transmitting voice message has possible stress.

3. Development of clear and correct conclusions for each separate part.
4. Provided that the conclusion is made about the unreliability of the received information, it is necessary to develop the procedure for confirming this information.
5. Then the entire sequence of results is analyzed by an expert.

It should be noted that when determining the unreliability of information caused by the stressful state of a helicopter pilot during a voice radio exchange, it is necessary to take into account the Lippold effect, which is that all human muscles, including the vocal cords, are subject to micro-fluctuations with the frequency of 8–12 Hz, the frequency of these fluctuations in stressful condition rises to 12 Hz.

It seems natural to introduce the tests in the control mode, i.e., the sets of input and relevant output data [16].

Formally, the test can be considered as the sequence of input vectors $\vec{X}_1, \ldots, \vec{Xn}$, that reflect the optimal (or reference) voice parameters and the corresponding output vectors $\vec{Y}_1, \ldots, \vec{Y}_N$, that reflect the actual voice parameters obtained via the communication channel, where N is the test length, i.e.

$$T = \left(\vec{X}_1, \ldots, \vec{X}_N; \vec{Y}_1, \ldots, \vec{Y}_N\right). \tag{4.54}$$

We assume that the test will detect the unreliability of information if the sequence of output vectors is different at some point from the original sequence $\vec{Y}_1, \ldots, \vec{Y}_N$.

We assign the set $\{s_i\}$ of tests (or states), each of which is characterized by a N_i-dimensional vector. We define the state estimate as the correlation $s_i \to \alpha_i$, where $i = 1, \ldots, m$.

We introduce the following verification principle:

(1) We introduce the system of reference estimates $\gamma_1, \ldots, \gamma_m$.
(2) Let us assign a sequence of numbers for each estimate β_i.
(3) If the condition in equation $(\alpha_i - \gamma_i) < \beta_i$ is satisfied, then we believe that the *i*th assessment did not reveal the unreliability of the information.
(4) If the condition in equation $(\alpha_i - \gamma_i) \geq \beta_i$ is satisfied, then we believe that the *i*th assessment revealed the unreliability of the information.

Then, the following coding principle is introduced:

Each *i*th assessment is assigned the number λ_i, which can take one of two possible values: 0—if the unreliability of information (due to external factors or strong emotional stress) is not detected, or 1—if these signs are detected.

We define the finite set $\{\pi_j\}$ of checks that have two outcomes: positive or negative. It seems natural to define π_j as the function of the whole set of estimates or $\pi_j = f\left(\alpha_1^j, \ldots, \alpha_m^j\right)$. We assume that if the check did not reveal the signs of unreliable information, then $\pi_j = 0$ and $\pi_j = 1$ otherwise.

If $\forall i\ \lambda_i^j = 0$, then $\pi_j = 0$. If $\forall i\ \lambda_i^j = 1$, then $\pi_j = 1$. In other cases, it is intended to check as follows [15, 17]:

We sort the estimates by priority and to each estimate we assign the weight $m_i m_i$ and calculate the value $\mathrm{M} = \sum_{i=1}^{m} m_i \lambda_i$. Then, we compare this value with a certain threshold value M′. If $\mathrm{M} < \mathrm{M}'$, then we believe that the unreliability of the information is not detected, and if $\mathrm{M} > \mathrm{M}'$, then we believe that it is detected.

In the process of detection of unreliability of information due to external factors or strong emotional stress, two types of errors are possible:

(1) The unreliability of the information in the *i*th assessment was not found at its presence. Let us indicate its probability ε_i:

$$\varepsilon_i = P(\lambda_i = 0/\lambda_i = 1); \tag{4.55}$$

(2) The unreliability of the information in the *i*th assessment was found at its absence. Let us indicate its probability δ_i:

$$\delta_i = P(\lambda_i = 1/\lambda_i = 0). \tag{4.56}$$

It is assumed that an error in the detection of unreliable information can occur for two reasons:

(1) An error due to received information;
(2) An error due to incorrect evaluation of the received information.

In the first case, various factors, such as meteorological conditions, equipment error, interference, terrain, etc., affect the unreliability of information.

We suppose that in the ith assessment t_i such factors where $i = 1, \ldots, m$ are taken into account, where we indicate the probability of each factor occurring as $\eta^i_{t_i}$. The probability of an error for the second case is indicated as $p(\Delta)$.

Then, according to the formula of total probability, given that different factors can influence each other, as well as errors due to the receipt of inaccurate information and incorrect assessment of reliable information can occur simultaneously, we have:

$$\varepsilon_i = \left[\left(\sum_{t_i=1}^{k_i} \eta^i_{t_i} P\left(\lambda_i = 0/\eta^i_{t_i}\right)\right) + p(\Delta)\right] - \left[p(\Delta) \sum_{t_i=1}^{k_i} \eta^i_{t_i} P\left(\lambda_i = 0/\eta^i_{t_i}\right)\right] \tag{4.57}$$

Similarly,

$$\delta_i = \left[\left(\sum_{t_i=1}^{k_i} \eta^i_{t_i} P\left(\lambda_i = 0/\eta^i_{t_i}\right)\right) + p(\Delta)\right] - \left[p(\Delta) \sum_{t_i=1}^{k_i} \eta^i_{i_i} P\left(\lambda_i = 0/\eta^i_{t_i}\right)\right]. \tag{4.58}$$

It seems natural to conclude that in order to calculate the probability of error during the entire test, and not with respect to individual assessments, it is necessary to minimize the number of factors affecting the error.

In this section, the attempt was made to analyze the problem connected with the navigation safety of helicopter flights under current conditions. Since voice information exchange is today practically the only means of ensuring the reliability of navigation support during the flight of helicopters along the route, the accuracy of information transmission by this method is the main condition for ensuring the safety of helicopter flights. The probability of error in determination of the reliability of information transmitted by voice message from the helicopter taking into account possible errors of the first and second type was obtained by the mathematical methods of probability theory and mathematical statistics.

However, it should be noted that the voice method of transmitting messages from the helicopter is not a sufficiently reliable method of transmitting information. Thus, the directions of further research in this area should be carried out in terms of obtaining new methods of navigation support of the helicopter.

References

1. Apollonov AA (2010) Improving the accuracy and reliability of user equipment in high latitudes when using pseudosatellites. Sci Bull MSTUCA (169)
2. Amirkhanov AV (2017) The issues of pairing of the system of instrument-technological modelling with the library of optimization algorithms. Amirkhanov AV, Glushko AA, Gladkikh AA, Makarchuk VV, Shahnov VA (eds). Nanoindustry 74:389–394
3. Baburov VI, Bestugin AR, Ivantsevich NV, Kirshina IA, Sauta OI, Filin AD, Shatrakov YG (2016) Comprehensive assessment of navigation information at a priori uncertainty. Issues Radio Electron 6:66–70
4. Zatuchny DA (2010) Assessment of the degree of influence of various factors on the navigation definitions of aircraft with the use of SRNSS. Sci Bull Moscow State Tech Univ Civil Aviation 159:143–147
5. Zatuchny DA (2018) To the problem of sustainability and radio safety of navigation systems of civil aircraft using precise satellite navigation systems. In: Proceedings of the international symposium Reliability and quality, vol 1, pp 38–40
6. Zatuchny DA (2010) Identification of false information on the voice of the pilot. In: Proceedings of the international symposium reliability and quality, vol 1, pp 429–430
7. About Global Navigation Satellite System GLONASS: The order of the President of the Russian Federation from 18.02.1999. № 38-rp. Garant. Information and legal support [Electronic resource]. https://base.garant.ru/180169/
8. Rubtsov VD (2015) The method of calculation of distributions of interference and its mixture with a signal using experimental curves of the distribution of the envelope of the interference. Inf Commun 2:57–61
9. Severtsev NA, Beckov AV (2009) System analysis of the theory of security. M.: Theis, p 452
10. Soloviev YA (2000) Satellite navigation systems. – M.: Eco-Trendz, p 270
11. Zatuchny DA, The assessment of the functional reliability of the combined means of communication of air traffic management systems and navigation support of aircraft based on satellite radio navigation systems: Thesis for the degree of candidate of Technical Sciences. 05.22.13 / Zatuchny Dmitry Aleksandrovich
12. Borsoev VA (2004) Methods and means of navigational flight support and aircraft control in high latitudes: Thesis for the degree of Doctor of Technical Sciences. Speciality 05.22.13 - Navigation and air traffic control. V.A. Borsoev – M.: MSTUCA, p 301

13. Sharov VD (2016) Flight safety risk management methodology at the airline level: Thesis for the degree of Doctor of Technical Sciences. Speciality 05.22.14 - Operation of air transport. – M.: MSTUCA, p 398
14. Baranov EV (2008) Improving the reliability of assessing the accuracy of navigation definitions for satellite navigation systems by the methods of semi-natural modelling Thesis for the degree of candidate of Technical Sciences. Speciality 05.22.13 - Navigation and air traffic control. E.V. Baranov – M.: MSTUCA, p 129
15. Sobolev SP (2007) Methods for assessing and monitoring of the integrity of the on-board navigation equipment of the satellite landing system: thesis for the degree of candidate of Technical Sciences: 05.12.14. - St. Petersburg, p 179
16. Revnivykh SG (2006) Development of software and algorithmic support of the monitoring system of global navigation satellite systems: thesis for the degree of candidate of Technical Sciences: 05.07.09. - Moscow, p 166
17. Filin AD, Shatrakov YG, Yakovlev VT, Yushkov AV (2017) The impact of reliability of flight support subsystems on the likelihood of occurrence of flight accidents. Sci Bull Military Ind Complex Russia 3:68–74

Annex A

Equipment Type Certificate No. 399 issued by the Interstate Aviation Committee.

O. I. Zavalishin et al., *Modern Requirements for Noise Immunity Aircraft Navigation Equipment*, Springer Aerospace Technology,
https://doi.org/10.1007/978-981-16-0073-9

МЕЖГОСУДАРСТВЕННЫЙ АВИАЦИОННЫЙ КОМИТЕТ
КОМИССИЯ ПО СЕРТИФИКАЦИИ АЭРОДРОМОВ И ОБОРУДОВАНИЯ

INTERSTATE AVIATION COMMITTEE
COMMISSION FOR AERODROME AND EQUIPMENT CERTIFICATION

СЕРТИФИКАТ ТИПА
TYPE CERTIFICATE
№ 399

НАСТОЯЩИЙ СЕРТИФИКАТ ВЫДАН
THIS CERTIFICATE IS ISSUED TO

Обществу с ограниченной ответственностью "НППФ СПЕКТР"
Россия, 123154, г. Москва, бульвар генерала Карбышева, дом 9, корпус 1
Limited liability company "NPPF SPECTR"
General Karbyshev Boulevard, 9, bld. 1, Moscow, 123154, Russia

И УДОСТОВЕРЯЕТ, ЧТО ТИПОВАЯ КОНСТРУКЦИЯ
AND CERTIFIES, THAT THE TYPE DESIGN OF

Локальной контрольно-корректирующей станции ЛККС-А-2000 КНТА.466539.005
с программным обеспечением в составе:
- **операционная система QNX4 V4.25**
- **прикладное программное обеспечение КНТА.00219.05**

Local Monitoring Corrective Station LKKS-A-2000 KHTA.466539.005
with software consisting of:
- *operating system QNX4 V4.25*
- *applications software KHTA.00219.05*

СООТВЕТСТВУЕТ СЕРТИФИКАЦИОННЫМ ТРЕБОВАНИЯМ.
COMPLIES WITH THE CERTIFICATION REQUIREMENTS.

ОГРАНИЧЕНИЯ ДЕЙСТВИЯ НАСТОЯЩЕГО СЕРТИФИКАТА СОДЕРЖАТСЯ В ПРИЛОЖЕНИИ, ЯВЛЯЮЩЕМСЯ НЕОТЪЕМЛЕМОЙ ЧАСТЬЮ СЕРТИФИКАТА.
CERTIFICATE LIMITATIONS ARE CONTAINED IN THE APPENDIX FORMING AN INTEGRAL PART OF THIS CERTIFICATE.

СЕРИЙНОЕ ПРОИЗВОДСТВО:
THE QUANTITY PRODUCTION AT
Общество с ограниченной ответственностью "НППФ СПЕКТР"
Россия, 123154, г. Москва, бульвар генерала Карбышева, дом 9, корпус 1.
Limited liability company "NPPF SPECTR"
General Karbyshev Boulevard, 9, bld. 1, Moscow, 123154, Russia.

02.12.2005
МОСКВА
MOSCOW

В.Н.Гончаров
Председатель
Комиссии МАК
V.N. Goncharov
Chairman
of the IAC Commission

Annex B [69]

Title page of the acceptance test report of the complex automated system for collecting and communicating to aviation users in the airspace of the Russian Federation of information on the status of orbital groupings of the global navigation satellite system and means of functional augmentations.

APPROVE	AGREED
Deputy Minister of Transport of Russian Federation	Deputy Head of Federal Air Transport Agency
/signature/ A.N. Nedosekov	/signature/ D.V. Savitsky
« » 2011	« » 2011

AGREED	AGREED
First Deputy General Designer	General director LLC RDPF Spektr
Deputy General Director of Concern of Air Defense "Almaz-Antey"	
/ signature / P.A. Sozinov	/ signature / O.I. Zavalishin
________ 2011	________ 2011

Copy No.__________

O. I. Zavalishin et al., *Modern Requirements for Noise Immunity Aircraft Navigation Equipment*, Springer Aerospace Technology,
https://doi.org/10.1007/978-981-16-0073-9

CERTIFICATE of acceptance tests of the integrated automated system for collecting and communicating to aviation users in the airspace of the Russian Federation of information on the status of orbital groupings of the global navigation satellite system and means of functional extensions. KNTA.466539.009, developed by LLC RDPF Spektr which passed preliminary tests 2011.

Annex C

The order of the Federal Air Transport Agency No. AU-142-r "On the acceptance for equipping the ground local monitoring and control station of the GLONASS/GPS combined navigation satellite system LMCS-A-2000".

MINISTRY OF TRANSPORT OF THE RUSSIAN FEDERATION

(MINTRANS OF RUSSIA)

FEDERAL AIR TRANSPORT AGENCY

(ROSAVIATSIYA)

ORDER

May 19, 2006 Moscow No. AU-142-r

On the acceptance for equipment of the ground local monitoring and control station of the combined navigation satellite system GLONASS/GPS LMCS-A-2000

In accordance with the "Terms of Reference for the development of the ground local monitoring and control station (LMCS)", approved by the Deputy Head of the State Regulation Department of Air Traffic Management of the State Civil Aviation Service of the Ministry of Transport of the Russian Federation, on February 15, 2002, the LLC RDPF Spektr developed and manufactured the ground local monitoring

O. I. Zavalishin et al., *Modern Requirements for Noise Immunity Aircraft Navigation Equipment*, Springer Aerospace Technology,
https://doi.org/10.1007/978-981-16-0073-9

and control station LMCS-A-2000 intended for use as part of radio engineering equipment of airfields as a sensor of differential corrections, service information and control of integrity of GLONASS/GPS at the use of LMCS-A-2000 as the facility for approach on the 1 ICAO category, navigation and as a part of the terrestrial station of broadcast the automatic dependent surveillance (ADS-B).

LMCS-A-2000 has passed acceptance and operational tests and has the type certificate No. 399 from 02.12. 2005, issued by the Interstate Aviation Committee.

In order to improve flight safety and air traffic management efficiency in the centers of the unified air traffic management system, I propose:

1. To Give the Heads of State Enterprises and Civil Aviation Organizations an Access to Equipment of LMCS-A-2000;
2. To the Director of the Federal State Unitary Enterprise "State Research Institute of Air Navigation" to ensure the provision of methodological assistance to state enterprises and civil aviation organizations of Russia when commissioning the LMCS-A-2000.
3. To Assign Control Over the Implementation of This Order to the Deputy Head of the Federal Air Transport Agency, D. V. Odintsov.

Acting Head / signature / N. N. Zlenko.

Annex D

The order of FSUE "State ATM Corporation" No. 049 from 04.02.2014 "On the commissioning of the equipment of the Monitoring Centre in the General Directorate of FSUE ATM State Corporation".

Federal Air Transport Agency

(ROSAVIATSIYA)

Federal State Unitary Enterprise

"STATE CORPORATION FOR THE ORGANIZATION OF AIR MOVEMENT IN THE RUSSIAN FEDERATION"

(FSUE "State ATM Corporation")

ORDER

February 04, 2014 Moscow No. 049

About the commissioning of the equipment of the Monitoring Centre in the General Directorate of FSUE "State ATM Corporation"

The equipment of the Centre for collecting and communicating to aviation users in the RF of information about monitoring of the status of the orbital groupings of the GLONASS/GPS systems (hereinafter—CAS CCIM) KNTA 466,539.009–001, manufactured by LLC RDPF Spektr, passed the acceptance and operational

O. I. Zavalishin et al., *Modern Requirements for Noise Immunity Aircraft Navigation Equipment*, Springer Aerospace Technology,
https://doi.org/10.1007/978-981-16-0073-9

tests in the general directorate of the Federal State Unitary Enterprise "State ATM Corporation" and in accordance with the Certificate on the results of operational tests is recommended for commissioning.

For the organization of the CAS CCIM operation in the general directorate of FSUE "State ATM Corporation"

I Hereby Order:

1. To put into operation from February 10, 2014, the equipment of the Monitoring Centre (KNTA 466,539.009–001) in the general directorate of the FSUE "State ATM Corporation". Prior to the commissioning of the regional monitoring centers of CAS CCIM, to carry out the operation of the CAS CCIM under the supervision of the developer.
2. To the Chief of CBD A. F. Kolotukhin to Organize the Use of CAS CCIM for Its Intended Purpose.
3. To the Information Technology Director V. N. Arkhipov to organize and ensure round-the-clock technical operation of the equipment of the CAS CCIM in accordance with the requirements of the operational technical documentation.
4. To the Director of the implementation of the federal program and modernization K. K. Kaplya together with the Director of Information Technologies V. N. Arkhipov with the further development of CAS CCIM and the introduction of regional central and local stations (LCS, MLCS, LMCS) to organize the necessary revision of the main CAS CCIM.
5. To the ATM and ASM Director I. V. Sitnikov, together with the head of CBD A. F. Kolotukhin to organize work on the revision and maintenance of the web-site of CAS CCIM.
6. To the Chief Accountant N. N. Gogopulo to accept to the balance of the established procedure the CAS CCIM equipment with the useful life of 7 years.
7. To entrust the coordination of work and control over the execution of this order to the deputy director general of the FSUE "State ATM Corporation" S. N. Pogrebnov.

Acting Director General / Signed / V. R. Gulchenko.

Annex E

The Certificate of the Federal Agency of the Russian Federation for Technical Regulation and Metrology (Rosstandart) on the Type of Measuring Instrument RU.C.27.002.A No. 51176.

FEDERAL AGENCY

ON TECHNICAL REGULATION AND METROLOGY

CERTIFICATE

on approval of the type of measuring instruments

RU.C.27.002.AN2 51176

Valid till June 24, 2018.
NAME OF TYPE OF MEASUREMENT INSTRUMENTS.
Ground and flight test complexes (GFTC) KNTA.466539.014.
MANUFACTURER.
Limited Liability Company "RDPF Spektr", Moscow.
REGISTRATION No. 53879–13.
DOCUMENT FOR TEST OF.
KNTA.466539.014.

O. I. Zavalishin et al., *Modern Requirements for Noise Immunity Aircraft Navigation Equipment*, Springer Aerospace Technology,
https://doi.org/10.1007/978-981-16-0073-9

INTERVAL BETWEEN TESTS 1 year.

The type of measuring instruments is approved by the order of the Federal Agency for Technical Regulation and Metrology from June 24, 2013 No. 610.

The description of the type of measuring instrument is a mandatory annex to this certificate.

Deputy Head of F.V. Bulygin the Federal Agency 28.06.2013 Series SI No. **010251**.

Valid till June 01, 2023.

Prolonged by the order of the Federal Agency for Technical Regulation and Metrology from June 01, 2018 No. 1093.

Deputy Head of the Federal Agency/signature/S. S. Golubev 13.06.2018

Annex F [1]

Title page of Act No. 163–11-IX on the results of the ground-based airfield and flight tests of GBAS satellite navigation and landing equipment prototypes of the II and III categories and the GBAS system elements on the IL-76 and IL-103 laboratory airplanes under conditions of actual interference of the airfield.

Gromov Flight Research Institute

APPROVE

First Deputy Head of FSUE "Gromov FRI" on science

__/ signature/ V.V. Tsyplakov

_______10_____2011

CERTIFICATE No. 163–11-1X.

On the results of the ground-based airfield and flight tests of GBAS satellite navigation and landing equipment prototypes of the II and III categories and the system elements on the IL-76 and IL-103 laboratory airplanes under conditions of actual interference of the airfield 2011.

O. I. Zavalishin et al., *Modern Requirements for Noise Immunity Aircraft Navigation Equipment*, Springer Aerospace Technology,
https://doi.org/10.1007/978-981-16-0073-9

Annex G

The extract from the flight test report of the LMCS-A-2000 at Sheremetyevo Airport with the conclusion on the possibility of commissioning.

FEDERAL AIR TRANSPORT AGENCY

(ROSAVIATSIYA)

FSUE "State ATM Corporation"

Branch "Moscow ATC Centre"

Sheremetyevo ATM Centre

SUBMIT FOR APPROVAL	APPROVE
Deputy Director for RTOP and AS of the branch "Moscow ATC Centre"	Director of the branch "Moscow ATC Centre"
FSUE "State ATM Corporation"	FSUE "State ATM Corporation"
/ signature / V.V. Krivosheev	/ signature / V.V. Uzhakov
"14" 08 2014	/illegible/ 2014

CERTIFICATE of flight test of the local monitoring and control station LMCS-A-2000 at Sheremetyevo Airport.

During the period from August 9, 2014 to August 11, 2014, by the crew of the AFL Beechcraft King Air 350i AC RA No. 02800, equipped with AF ADAPIS-112

O. I. Zavalishin et al., *Modern Requirements for Noise Immunity Aircraft Navigation Equipment*, Springer Aerospace Technology,
https://doi.org/10.1007/978-981-16-0073-9

pr. No. 002, of Aerocontrol branch FSUE State ATM Corporation carried out flight test on commissioning of the local monitoring and control station LMCS-A-2000 pr. No. 1013494.

Flight test was carried out by: AC commander On-board engineer-operator RLG leading engineer Head of OiPR OOS department	Measurements of parameters and characteristics of the LMCS-A-2000 were carried out in accordance with the requirements of the Federal Aviation Regulations "Flight tests of ground-based radio equipment for flight operations, aviation telecommunications and systems of lighting signal equipment of civil aviation" approved by the Order No. 1 of the Ministry of Transport of Russia (registered on March 10, 2005 by the Ministry of Justice of Russia, registration No. 6383), as amended by the Order No. 117 of the Ministry of Transport of the Russian Federation on April 20, 2011 (registered by the Ministry of Justice of Russia on June 21, 2011, registration No. 21092), and the order of the Ministry of Transport of Russia from June 29, 2012 No. MS-82-r. The measurement results are shown in the table of the annex to the flight test certificate of the LMCS-A-2000. **CONCLUSION** **The local monitoring and control station LMCS-A-2000 pr. No. 1013494 at Sheremetyevo Airport meets the operational requirements and is suitable for performance of instrumental landing approach with right MKp-066°, left MKp-066 °, right MKp-2460, left MKp-2460 on the 1 category of ICAO and can be operated without restrictions.**

Annex H

The extract from the flight test report of the LMCS-A-2000 at Domodedovo Airport with the conclusion on the possibility of commissioning.

FEDERAL AIR TRANSPORT AGENCY

(ROSAVIATSIYA)

FSUE "State ATM Corporation"

Branch "Moscow ATC Centre"

Domodedovo ATM Centre

SUBMIT FOR APPROVAL	APPROVE
Acting Head of the Domodedovo ATM Centre	Deputy Director for RTOP and AS of the branch "Moscow ATC Centre" FSUE "State ATM Corporation"
/ signature / V.A. Sitnikov	/ signature / V.V. Krivosheev
April 11, 2016	12.04.2016

CERTIFICATE of flight test of the local monitoring and control station LMCS-A-2000 at Domodedovo Airport.

During the period from March 4, 2016 to April 9, 2016, the crew of the AC Beechcraft King Air 350i AC RA No. 02800 and the crew of the AC Beechcraft King Air 350iBort RA No. 02814 equipped with flight monitoring equipment AD-AFIS- 112 pr. No. 002 and No. 003 Aerocontrol branch FSUE State ATM Corporation carried out flight test on commissioning of the local monitoring and control station LMCS-A-2000 pr. No. 013495.

O. I. Zavalishin et al., *Modern Requirements for Noise Immunity Aircraft Navigation Equipment*, Springer Aerospace Technology,
https://doi.org/10.1007/978-981-16-0073-9

Flight test was carried out by:

AC commander

On-board engineer-operator

KDP leading engineer

Chief Specialist of LLC RDPF Spektr

Measurements of parameters and characteristics of the LMCS-A-2000 were carried out in accordance with the requirements of the Federal Aviation Regulations "Flight tests of ground-based radio equipment for flight operations, aviation telecommunications and systems of lighting signal equipment of civil aviation" approved by the Order No. 1 of the Ministry of Transport of Russia (registered on March 10, 2005 by the Ministry of Justice of Russia, registration No. 6383), as amended by the Order No. 117 of the Ministry of Transport of the Russian Federation on April 20, 2011 (registered by the Ministry of Justice of Russia on June 21, 2011, registration No. 21092), and the order of the Ministry of Transport of Russia from June 29, 2012 No. MS-82-r.

The measurement results are shown in the table of the annex to the flight test certificate of the LMCS-A-2000.

CONCLUSION

The local monitoring and control station LMCS-A-2000 pr. No. 1013495 at Domodedovo Airport meets the operational requirements and is suitable for instrumental approach for aircraft landing with Mkp-136° R, Mkp-136 ° L and with Mkp-316 ° R, Mkp-316° L on the 1 ICAO category.

Annex I

The extract from the flight test report of the LMCS-A-2000 at Vnukovo Airport with the conclusion on the possibility of commissioning.

FEDERAL AIR TRANSPORT AGENCY

(ROSAVIATSIYA)

FSUE "State ATM Corporation"

Branch "Moscow ATC Centre"

Vnukovo ATM Centre

SUBMIT FOR APPROVAL	APPROVE
Chief of	Deputy Director of the branch
Vnukovo ATM Centre	"Moscow ATC Centre"
	FSUE "State ATM Corporation" RTOP and AES
/ signature/ G.D. Yarmush	/ signature / V.V. Krivosheev
October 27, 2015	/illegible/ 2015

CERTIFICATE of flight test of the local monitoring and control station LMCS-A-2000 at Vnukovo Airport.

During the period from October 22, 2015 to October 27, 2015, the crew of AC Beechcraft King Air 350i AC No. RA - 02,800, equipped with the ALA ADAFIS-112 pr. No. 003, of Aerocontrol branch FSUE State ATM Corporation carried out flight test on commissioning of the local monitoring and control station LMCS-A-2000 pr. No. 1013441.

O. I. Zavalishin et al., *Modern Requirements for Noise Immunity Aircraft Navigation Equipment*, Springer Aerospace Technology,
https://doi.org/10.1007/978-981-16-0073-9

Flight test was carried out by AC commander On-board engineer-operator RMG leading engineer Head of the department of LLC RDPF Spektr	Measurements of parameters and characteristics of the local monitoring and control station LMCS-A-2000 were carried out in accordance with the requirements of the Federal Aviation Regulations "Flight tests of ground-based radio equipment for flight operations, aviation telecommunications and systems of lighting signal equipment of civil aviation" approved by the Order No. 1 of the Ministry of Transport of Russia (registered on March 10, 2005 by the Ministry of Justice of Russia, registration No. 6383), as amended by the Order No. 117 of the Ministry of Transport of the Russian Federation on April 20, 2011 (registered by the Ministry of Justice of Russia on June 21, 2011, registration No. 21092), and the order of the Ministry of Transport of Russia from June 29, 2012 No. MS-82-r. The measurement results are shown in the table of the annex to the flight test certificate of the local monitoring and control station LMCS-A-2000. **CONCLUSION** **The local monitoring and control station LMCS-A-2000 pr. No. 1013441 at Vnukovo Airport meets the operational requirements and is suitable for performance of instrumental landing approach with MKp-013°, MKp-193°, MKp-058°, MKp-238° on the 1 category of ICAO and can be operated without restrictions.**

Annex J

The extract from the flight test report of the LMCS-A-2000 at Pulkovo Airport with the conclusion on the possibility of commissioning.

FEDERAL AIR TRANSPORT AGENCY

(ROSAVIATSIYA)

FSUE "State ATM Corporation"

Branch "North-West Air Navigation"

St. Petersburg ATM Centre

SUBMIT FOR APPROVAL	APPROVE
Deputy Head of the ATM Centre - Head of the ERTOS Service of St. Petersburg ATM Centre	Chief Of St. Petersburg ATM Centre of the Branch "North-West Air Navigation"
Yu. I. Zak	S.N. Kuzmin
15 December 2016	15 December 2016

CERTIFICATE of flight test of the local monitoring and control station LMCS-A-2000 at Saint-Petersburg Airport (Pulkovo).

During the period from December 13, 2016 to December 14, 2016, the crew of the AC An - 24 board. No. 46395, equipped with flight control equipment ASLK - 2005 pr. No. 0901, of Aerocontrol branch FSUE State ATM Corporation carried out special flight test on commissioning of the local monitoring and control station LMCS-A-2000 pr. No. 1013437.

O. I. Zavalishin et al., *Modern Requirements for Noise Immunity Aircraft Navigation Equipment*, Springer Aerospace Technology,
https://doi.org/10.1007/978-981-16-0073-9

Flight test was carried out by: AC commander Onboard engineer - operator Chief Specialist of LLC RDPF Spektr Leading Engineer	Measurements of parameters and characteristics of the LMCS-A-2000 were carried out in accordance with the requirements of the Federal Aviation Regulations "Flight tests of ground-based radio equipment for flight operations, aviation telecommunications and systems of lighting signal equipment of civil aviation" approved by the Order No. 1 of the Ministry of Transport of Russia (registered on March 10, 2005 by the Ministry of Justice of Russia, registration No. 6383), as amended by the Order No. 117 of the Ministry of Transport of the Russian Federation on April 20, 2011 (registered by the Ministry of Justice of Russia on June 21, 2011, registration No. 20), and the order of the Ministry of Transport of Russia from June 29, 2012 No. MS-82-r. in accordance with the point 1.1 of the "Plan and Time Schedule of Main Organizational and Technical Measures on the Transition to QNH near the St. Petersburg Airfield (Pulkovo)" and in connection with the mass development in the St. Petersburg Airfield (Pulkovo). The measurement results are shown in the table of the annex to the flight test certificate of the local monitoring and control station. **CONCLUSION** The local monitoring and control station LMCS-A-2000 pr. No. 1013437 at the St. Petersburg (Pulkovo) airfield meets the performance requirements and is suitable for performing an instrumental landing approach with MKp-096R°, MK-276° L, MKP-096° L and MKN-276 ° R in the 1 category of ICAO, and can be operated without restrictions.

Annex K

Information on implementation (the letter of FSUE "State ATM Corporation");

Federal Air Transport Agency

(ROSAVIATSIYA)

Federal State Unitary Enterprise

"STATE CORPORATION FOR THE ORGANIZATION OF AIR MOVEMENT IN THE RUSSIAN FEDERATION"

FSUE -State ATM Corporation

INFORMATION ON IMPLEMENTATION of the satellite navigation and landing system of the ICAO 1 category at airports in the Russian Federation and the system for monitoring and communicating to aviation users of information on the state of GNSS navigation fields in the interests of Civil Aviation in the Russian Federation with the direct participation of Zavalishin O. I., the applicant for the nomination "For a special personal contribution to the development of air navigation in Russia" of the G.N. Pirogov Air Navigation Development Fund.

Developed with the direct participation of the applicant—the Chief Designer and General Director of RDPF Spektr, Oleg Ivanovich Zavalishin in accordance with the ICAO standards for GBAS and Russian regulatory documents—the satellite navigation and precision landing approach system, the local monitoring and control station LMCS-A-2000 (functional extension to GNSS—GBAS) has successfully

O. I. Zavalishin et al., *Modern Requirements for Noise Immunity Aircraft Navigation Equipment*, Springer Aerospace Technology,
https://doi.org/10.1007/978-981-16-0073-9

passed all types of tests, is certified and in accordance with the federal target program "GLONASS", by the Decree of the President of the Russian Federation from May 17, 2007 No. 638 "On the use of the global GLONASS satellite system in the interests of the socio-economic development of the Russian Federation" was established, flown around the installation site and implemented in more than 100 airports of the Russian Federation as an instrument of navigation, monitoring of GNSS signals and precision instrumental approach.

The Federal Air Transport Agency issued the Order No. AU-142-r "On the acceptance for equipping of the ground local monitoring and control station of the combined navigation satellite system GLONASS/GPS LMCS-A-2000" at civil aviation airports.

The significant implementation projects are the installation and commissioning of the LMCS-A-2000 at Sheremetyevo, Domodedovo, Pulkovo, Sochi, Grozny airports, on Kunashir island, at the Krainy airport at Baikonur, etc.

A bright example of the implementation and demonstration of the advantages of the LMCS-A-2000 satellite navigation and landing system is also its successful use in the most difficult operating conditions in Antarctica to ensure ocean flight, approach and precision landing approach of the Il-76 aircraft to the Novolazarevskaya ice airfield over 10 seasons. This is a unique practical experience that has no analogues in the world.

The scale of the project from the development of standards to the implementation of the air navigation system in enterprises is unprecedented and has no analogues in the modern history of aviation and air navigation in Russia.

Considerable work on meeting the requirements of the ICAO standards and ensuring compatibility with foreign aircraft was carried out in conjunction with Eurocontrol.

With the direct participation of the applicant, the work was carried out on equipping domestic aircraft with GLS (GBAS Landing System) equipment and testing on the use of landing technology. 24 aircraft of the aviation enterprise Gazprom Aviation were equipped with precision approach systems.

The first landing on the LMCS-A-2000 system of a passenger Boeing of a Russian airline was carried out in collaboration with the airline Globus (S7 Airlines) with the direct participation of the chief navigator of Globus Ltd. Chuvikovsky Konstantin Valerievich at Roshino airport (Tyumen).

The equipment of LMCS-A-2000 is the equipment for radio engineering of airfield flights to ensure the procedures of zonal navigation, nonprecision and precision landing approach on GLONASS/GPS on the I (optional II) category of ICAO. At present, only the Honeywell firm (USA) has certified the same equipment in the world, which, unlike the domestic GBAS (LMCS), works only on the GPS constellation.

Installation of one LMCS-A-2000 for equipped with GLS on-board equipment aircraft allows the construction of flight patterns along a curvilinear trajectory, nonprecision and precision landing approaches from both ends of all FLS of the airfield of the I (optional II category) category of ICAO, which saves significant funds for the air navigation system for installation, flight tests and operation of area navigation and landing systems.

The LMCS ground equipment monitors the integrity and quality (SQM) of satellite navigation signals, notifying the user of GLONAS/GPS satellite numbers that are distorted by the satellites themselves, readiness prediction (visibility of satellites in the LMCS area).

Functional extension to ground-based GNSS station LMCS-A-2000 transmits the integrity monitoring parameters of satellite navigation signals, the landing data block to Aircraft simultaneously for all the airport's approach procedures, differential corrections to satellite pseudo-range and other navigation information to ATM and RTO, information on the status of GLONAS/GPS satellite constellations and station equipment, as well as information on the possibility/impossibility of performing a typical operation in the airfield area on GNSS in differential and autonomous modes. Navigation positioning of Aircraft in the GLONASS/GPS differential mode on the LMCS is performed with meter accuracy.

Piloting during approaches on the LMCS-A-2000 is aligned with flights using traditional means and does not require additional skills from the crews. The display is made on standard devices. Information on the status of the GLONASS/GPS satellite constellations and the LMCS-A-2000 equipment is displayed on the ATM and RTO control panels.

Approaching on LMCS reduces flight time, reduces the amount of goings to the second-round due to meteorological conditions, increases airfield throughput, and reduces fuel consumption and the negative impact on the environment in the airfield area with minimal operating costs.

Installation of the system does not require construction work, takes a short time (4–6 days), does not require periodic seasonal fly-overs, has minimal operating costs and works automatically without operators.

In a number of remote, hard-to-reach, mountain, border, northern, island and far-eastern regions, LMCS-A-2000 is the only instrument of provision of an instrumental approach, especially in adverse weather conditions.

As part of the Federal Target Program "GLONASS", the mobile option for the LMCS-A-2000 was developed for the Ministry of Transport of the Russian Federation. It takes several hours to deploy it, which is very important in emergency situations, special operations of law enforcement agencies and in the provision of emergency medical care.

The applicant for the nomination—O. I. Zavalishin is the chief designer of complex automated system for monitoring and communicating to aviation users of information on the state of GNSS navigation fields in the interests of Civil Aviation (CAS CCIM). The CAS CCIM system (main and regional GNSS monitoring centers), developed with the direct participation of the applicant—Zavalishin Oleg Ivanovich (chief designer and general director of LLC RDPF Spektr) in accordance with ICAO standards on GNSS and Russian regulatory documents, successfully passed all types of tests.

The CAS CCIM system (GNSS monitoring centers) is installed and put into operation in the FSUE "State ATM Corporation" and 12 branches of the FSUE "State

ATM Corporation" in the cities—Moscow, Samara, Rostov-on-Don, Tyumen, Yekaterinburg, Yakutsk, Magadan, St. Petersburg, Novosibirsk, Irkutsk, Krasnoyarsk, Khabarovsk, Vnukovo a/f as an instrument of monitoring of GNSS signals.

The CAS CCIM system developed and put into operation meets the requirements of the ICAO GNSS standards (Annex 10, Volume I, paragraph 2.1.4.2 "The State authorizing flight operations based on the use of GNSS should ensure the registration of GNSS data connected with these flights").

The CAS CCIM system provides aviation users with information on the archive, current (real-time) and predicted state of the signals of the Global Navigation Satellite System (GNSS), to make decisions about the possibility/prohibition of aircraft to perform typical flight procedures, taking into account the planned routes, stages and time intervals of flights in differential and autonomous modes of on-board GNSS receivers.

The CAS CCIM system provides registration of information on the GNSS status related to specific flights that are necessary in the investigation of flight incidents, ensures the issuance of NOTAMs in the event that the RAIM function is unavailable, which does not allow flights in the absence of the required satellite geometry to a set point. The scope of the CAS CCIM system is ATM, Civil Aviation airports, airlines, navigator services and other users.

Director on Implementation of Federal Targeted Programs and Modernization of FSUE "State ATM Corporation"/ signature /K. K. Kaplya.

Annex L

The program for detection of “false satellites” based on the integration of satellite and inertial navigation systems.

Language C++

```
#include <iostream>
#include <cmath>
using namespace std;
int main()
{
        setlocale(LC_ALL, "");
        double c = 300000000;
        int kol;
        double
                D;
        cout << "Enter the number of navigation satellites ";
        cin >> kol;
        double sam[4];
        double **mass = new double*[kol];
        for (int i = 0; i < kol; i++) {
                mass[i] = new double[4];
        }
        cout << "Enter the coordinates and the time scales, " << kol << "
satellite (s) in the form of (x y z T)" << endl;
```


O. I. Zavalishin et al., *Modern Requirements for Noise Immunity Aircraft Navigation Equipment*, Springer Aerospace Technology,
https://doi.org/10.1007/978-981-16-0073-9

```
for (int i = 0; i < kol; i++) {
        cout << i + 1 << " Navigation satellite:" << endl;
        for (int j = 0; j < 4; j++) {
                cin >> mass[i][j];
        }
}
cout << "Enter the coordinates of the aircraft and the discrepancy of
the time scales, in the form (x y z T)" << endl;
cin >> sam[0]; cin >> sam[1]; cin >> sam[2]; cin >> sam[3];
double d[4];
for (int i = 0; i < kol; i++)
{
        d[i] = sqrt(pow(mass[i][0] - sam[0], 2) + pow(mass[i][1] - sam[1],
2) + pow(mass[i][2] - sam[2], 2)) + c * (mass[i][3] - sam[3]);
}
cout << "Enter the coordinates and the discrepancy of time scales
with IIS, in the form (x y z T) " << endl;
double iis[4];
cin >> iis[0] >> iis[1] >> iis[2] >> iis[3];
D = 0;
double d_iis = sqrt(pow(iis[0] - sam[0], 2) + pow(iis[1] - sam[1], 2) +
pow(iis[2] - sam[2], 2)) + c * (iis[3] - sam[3]);
for (int i = 0; i < kol; i++)
        D = D + d[i];
D = D / kol;

double X;
cout << "Enter maximum inaccuracy: ";
cin >> X;
if (d_iis - D >= X)
        cout << "The aircraft system does not work accurately!";
else cout << "The aircraft system does work accurately!";
system("pause");
return 0;
}
```

The manufacturer's authorised representative in the EU is Springer Nature Customer Service Centre GmbH, Europaplatz 3, 69115 Heidelberg, Germany. If you have any concerns regarding our products, please contact ProductSafety@springernature.com

Printed and bound by CPI Group (UK) Ltd, Croydon, CR0 4YY
30/09/2026
02225294-0002